Multilevel Modeling
Methodological Advances, Issues, and Applications

Multivariate Applications Book Series

Sponsored by the Society of Multivariate Experimental Psychology, the goal of this series is to apply complex statistical methods to significant social or behavioral issues, in such a way so as to be accessible to a nontechnical-oriented readership (e.g., nonmethodological researchers, teachers, students, government personnel, practitioners, and other professionals). Applications from a variety of disciplines, such as psychology, public health, sociology, education, and business, are welcome. Books can be single or multiple authored or edited volumes that: (1) demonstrate the application of a variety of multivariate methods to a single, major area of research; (2) describe a multivariate procedure or framework that could be applied to a number of research areas; or (3) present a variety of perspectives on a controversial topic of interest to applied multivariate researchers.

There are currently eight books in the series:

- *What if There Were No Significance Tests?* co-edited by Lisa L. Harlow, Stanley A. Mulaik, and James H. Steiger (1997).
- *Structural Equation Modeling With LISREL, PRELIS, and SIMPLIS: Basis Concepts, Applications, and Programming,* written by Barbara M. Byrne (1998).
- *Multivariate Applications in Substance Use Research: New Methods for New Questions,* co-edited by Jennifer S. Rose, Laurie Chassin, Clark C. Presson, and Steven J. Sherman (2000).
- *Item Response Theory for Psychologists,* co-authored by Susan E. Embretson and Steven P. Reise (2000).
- *Structural Equation Modeling With AMOS: Basic Concepts, Applications, and Programming,* written by Barbara M. Byrne (2001).
- *Conducting Meta-Analysis Using SAS,* written by Winfred Arthur, Jr., Winston Bennett, Jr., and Allen I. Huffcutt (2001).
- *Modeling Intraindividual Variability With Repeated Measures Data: Methods and Applications,* co-edited by D. S. Moskowitz and Scott L. Hershberger (2002).
- *Multilevel Modeling: Methodological Advances, Issues, and Applications,* co-edited by Steven P. Reise and Naihua Duan (2003).

Anyone wishing to submit a book proposal should send the following: (1) author/title, (2) timeline including completion date, (3) brief overview of the book's focus, including table of contents, and ideally a sample chapter (or more), (4) a brief description of competing publications, and (5) targeted audiences.

For more information please contact the series editor, Lisa Harlow, at: Department of Psychology, University of Rhode Island, 10 Chafee Road, Suite 8, Kingston, RI 02881-0808; Phone: (401) 874-4242; Fax: (401) 874-5562; or e-mail: LHarlow@uri.edu. Information may also be obtained from members of the advisory board: Leona Aiken or Steve West (both of Arizona State University), Gwyneth Boodoo (Educational Testing Service), Barbara M. Byrne (University of Ottawa), Scott E. Maxwell (University of Notre Dame), or David Rindskopf (City University of New York).

Multilevel Modeling
Methodological Advances, Issues, and Applications

Edited by

Steven P. Reise
Naihua Duan
University of California, Los Angeles

Routledge
Taylor & Francis Group

NEW YORK AND LONDON

First Published by
Lawrence Erlbaum Associates, Inc., Publishers
This edition published 2012 by Routledge
711 Third Avenue, New York, NY 10017
27 Church Road, Hove, East Sussex BN3 2FA

Routledge is an imprint of the Taylor & Francis Group, an informa business

Camera ready copy of the text for this volume was provided by the editors.

Cover design by Kathryn Houghtaling Lacey

Library of Congress Cataloging-in-Publication Data

Multilevel modeling : methodological advances, issues, and applications /
edited by Steven P. Reise, Naihua Duan.
    p. cm. – (Multivariate applications)
  Includes bibliographical references and index.
ISBN 0-8058-3670-5 (cloth : alk. paper)
ISBN 0-8058-5170-4 (pbk : alk. paper)
1. Multiple comparisons (Statistics) 2. Multivariate analysis. I. Reise, Steven Paul. II. Duan, Naihua, 1949- III. Multivariate applications book series.
QA278.4 .M78   2002
519.5'35—dc21   2001054343

Publisher's Note
The publisher has gone to great lengths to ensure the quality of this reprint
but points out that some imperfections in the original may be apparent.

Contents

v

Preface

Multilevel modeling is currently a very active research area with statistical advances occurring at a rapid pace. It is also being increasingly used in the applied world of data analysis because these techniques can address issues that are problematic with traditional approaches. In this edited volume, many of the leading multilevel modeling researchers from around the world illustrate their current work. These authors were invited to write chapters based on an open competition and call for papers in 1999. All chapters were peer reviewed by us and at least one external reviewer. We thank, Steven West, Dougal Hutchinson, Antony Fielding, Russell Ecob, Peter Bentler, Mike Seltzer, Juwon Song, Robert Cudeck, and Albert Satorra for reviewing one or more chapters.

As per the title, the chapters focus on new statistical advances (e.g., Cudeck & du Toit's chapter on nonlinear models for repeated measures data), methodological issues (e.g., Seltzer & Choi's chapter on outlier detection), and current applications of multilevel modeling (e.g., Baumler, Harrist, & Carvajal's chapter that illustrates analyses from the safer choices project). Because most chapters address each of these issues, it was impossible to separate them into distinct and coherent sections. Instead, the chapters are ordered in the sequence in which we were able to edit them. Thus, the chapter ordering in no way reflects our subjective judgments of quality. Our thanks is extended to James M. Henson for translating all manuscripts into camera-ready copy.

We believe that this volume will be most beneficial for researchers with advanced statistical training and extensive experience in applying multilevel models. Several chapters are quite statistically advanced (e.g., Cudeck & du Toit, Bentler & Liang, Fielding), although applications of these new techniques to real data are often provided. This book is probably not an optimal choice as an introductory graduate–level text, but may serve as a supplement to such a text. At the end of this volume, we provide a list of author contact information.

1

Nonlinear Multilevel Models for Repeated Measures Data

Robert Cudeck
University of Minnesota

Stephen H. C. Du Toit
Scientific Software International, Inc.

Hierarchical or multilevel models (e.g., Bryk & Raudenbush, 1992; Goldstein, 1987) have become valuable in many research domains as a way to account for naturally occurring or experimentally imposed hierarchical sampling configurations. The classic example in education is the nesting structure that arises when students are sampled from classrooms, which in turn are sampled from schools. Several desirable features are associated with multilevel models for hierarchical designs, including improved efficiency of estimation and more sensitive evaluation of treatment effects.

An important application of multilevel models is the description of individual change in repeated measures experiments or longitudinal studies. The technology in this situation is essentially the same as in the study of treatment effects in a hierarchically structured sample, but with different objectives. The goal is to summarize the average change in the population over time, while also describing individual patterns of development. Individual observation vectors may well be unique. Typically, they differ from the mean response as well. The statistical problem is to fit both the mean vector and the collection of individual responses by the same function, but with distinct parameterizations. A multilevel model for the repeated measures problem is often referred to as a *mixed-effects model,* after the decomposition of a response into a fixed term for the mean vector plus a term for individual variation.

Several algorithms have been presented for maximum likelihood estimation of the linear mixed-effects model assuming normality of the residuals and random effects. Software is widely available (e.g., Kreft, De Leeuw, & Van der Leeden, 1994). Estimation of the nonlinear model is more difficult, even though nonlinear models are typically more realistic. Consequently, although maximum likelihood estimation in linear models is relatively standard, its use with nonlinear models is uncommon. The problem is that the likelihood function for parameters of the nonlinear model involves an integration over the random effects to obtain the unconditional distribution of the response. Except in special cases, the integration cannot be carried out explicitly (Davidian & Giltinan, 1995, chap. 4). To estimate the model, at least three different strategies have been suggested, only the first of which produces exact maximum likelihood estimates; (a) empirical Bayes or fully Bayesian methods utilizing the EM algorithm or a data augmentation method such as the Gibbs sampler (Walker, 1996); (b) approximating the nonlinear response function by a first-order Taylor series, treating the resulting problem as a linear model that is handled with standard techniques (e.g., Lindstrom & Bates, 1990); (c) two-stage methods based on first obtaining estimates for each individual, then pooling information across subjects (Davidian & Giltinan, 1995, chap. 4; Vonesh, 1992; Vonesh & Chinchilli, 1997, chap. 7).

These methods often perform satisfactorily, certainly as general-purpose procedures. They work especially well in cases when the number of parameters in the function is small and the number of observations per individual is relatively large. All the same, limitations have been noted (Davidian and Giltinan, 1993; Roe, 1997). With the EM algorithm, convergence can be slow, irrespective of the method used to obtain moments of the missing data. With estimators based on linearization, the transformation of the nonlinear function involves an approximation whose accuracy is difficult to assess and that varies from one response function to another. With two-stage estimators, efficiency can be poor and the methods may not be applicable at all if the individual data are sparse. Consequently, results from both the approximate estimators and two-stage estimators can differ in nontrivial ways. An important justification for ongoing research on maximum likelihood estimation is that it provides a standard against which other methods may be evaluated.

In recent treatments, Davidian and Giltinan (1995) and Pinheiro and Bates (1995, sec. 2) suggested marginal maximum likelihood estimation of the nonlinear mixed-effects model (cf. Bock, 1989). Gauss-Hermite quadrature is recommended in both references for the integration over the random effects to obtain the marginal distribution of the response. However, no details have been presented about the way the likelihood function and the numerical approximation to it actually might be implemented. In their review of methods for approximating this likelihood function, Pinheiro and Bates (1995) reported numerical results for the Gaussian quadrature approach, but did not describe how the estimates summarized in their article were obtained. Pinheiro subsequently informed us (personal communication, December 10, 1998) that a general algorithm was employed for nonlinear function maximization from the S system (Chambers & Hastie, 1992, chap. 10). In their use of this method, they utilized an option for numerical differentiation. It is well known that this strategy is convenient and generally satisfactory for many problems (e.g., Dennis & Schnabel, 1983, sec. 5.6). On the other hand, it can be expensive in terms of function evaluations, and is subject to numerical inaccuracies when truncation or round-off errors swamp the calculations (Burden & Faires, 1993, chap. 4).

In this chapter, we also investigate marginal maximum likelihood estimation for the nonlinear mixed-effects model for continuous variables with structured covariance matrix for the residuals and a linear model for covariates at the second level. The model allows incomplete response vectors, data missing at random, time-varying covariates, and individual patterns of measurement. Similar to the approach of Pinheiro and Bates, the likelihood function is directly defined in terms of the quadrature formula used to approximate the marginal distribution. Estimates by the Newton-Raphson method will be obtained using explicit expressions for the gradient vector and numerical approximation to the Hessian matrix. This simple algorithm for the Hessian is generally twice as accurate as is a purely numerical method for second-order derivatives. Another benefit of this method is that the stochastic parameters can be easily computed from terms obtained for the gradient vector. The main advantages of the Newton-Raphson method is its reliable and rapid convergence rate and the fact that it provides an accurate approximation to the asymptotic covariance matrix of the estimates. An example with data from a repeated measures experiment and a second illustration with data from a longitudinal study are provided to demonstrate that the method performs satisfactorily.

THE NONLINEAR MIXED-EFFECTS MODEL

The nonlinear mixed-effects model is

$$\mathbf{y}_i = \mathbf{f}_i(\boldsymbol{\beta}_i, \mathbf{x}_i, \mathbf{z}_i) + \mathbf{e}_i$$

where $\mathbf{y}_i = (y_{i1}, \ldots, y_{in_i})'$ is the vector of n_i observed scores for the i-th individual,

$$\mathbf{f}_i(\boldsymbol{\beta}_i, \mathbf{x}_i, \mathbf{z}_i) = [f_{i1}(\boldsymbol{\beta}_i, \mathbf{x}_i, \mathbf{z}_i), \ldots, f_{i,n_i}(\boldsymbol{\beta}_i, \mathbf{x}_i, \mathbf{z}_i)]'$$

is a vector-valued, nonlinear function of coefficients, β_i, independent variables, $\mathbf{x}_i$, and covariate scores, $\mathbf{z}_i$. This notation for $\mathbf{f}_i$ is utilized because the number of observed scores for an individual, as well as the form of the function, may differ for each component of the response. The individual vectors, $\beta_i : (p \times 1)$, are the sum of fixed and random effects plus level-two covariates

$$\beta_i = \beta + \mathbf{B}\mathbf{b}_i + \mathbf{\Theta}\mathbf{Z}_i \qquad (1.1)$$

where $\beta : (p \times 1)$ are fixed parameters, $\mathbf{b}_i : (r \times 1)$, $r \leq p$, are random effects that vary across individuals. The second level regression coefficients are elements of the $(p \times v)$ matrix $\mathbf{\Theta}$. It is convenient subsequently to define $\theta = vec(\mathbf{\Theta})$. The design matrix $\mathbf{B} : (p \times r)$ usually has fixed values of unity or zero. It is used to associate particular elements of $\mathbf{b}_i$ with those of β, allowing for the possibility that some components of β_i do not vary, or that some element of $\mathbf{b}_i$ corresponds to two or more values of β, as is sometimes needed in the study of multiple populations. For example, if the first and third coefficients of β_i are random but the second is fixed with no level-two covariates, then

$$\mathbf{B} = \begin{pmatrix} 1 & 0 \\ 0 & 0 \\ 0 & 1 \end{pmatrix}$$

and $\beta_i = (\beta_1 + b_{i1}, \beta_2, \beta_3 + b_{i3})'$.

It is assumed that the residuals, $\mathbf{e}_i = \mathbf{y}_i - \mathbf{f}_i(\beta_i, \mathbf{x}_i, \mathbf{z}_i)$, have distribution $\mathbf{e}_i \sim \mathbf{N}(\mathbf{0}, \mathbf{\Lambda}_i)$ where the covariance matrix has structure

$$\mathbf{\Lambda}_i = \mathbf{\Lambda}_i(\boldsymbol{\lambda})$$

for parameters $\boldsymbol{\lambda} : (q_\lambda \times 1)$. The residuals often are taken to be independent with constant variance, σ^2, in which case $\mathbf{\Lambda}_i(\boldsymbol{\lambda}) = \sigma^2 \mathbf{I}$. In other situations, structures that specify either nonconstant variance or serial correlation or both can be accommodated. In still other cases, the error structure may depend on the mean response (e.g., Davidian & Giltinan, 1995, sec. 4.2.2). Although the order of $\mathbf{\Lambda}_i$ is n_i, the matrix depends on i only through its dimension.

The distribution of the random effects is assumed to be normal with expected value zero and covariance matrix $\mathbf{\Phi}$: $\mathbf{b}_i \sim \mathbf{N}(\mathbf{0}, \mathbf{\Phi})$. The covariance matrix is parameterized in terms of its Choleski factorization,

$$\mathbf{\Phi} = \mathbf{T}\mathbf{T}' \qquad (1.2)$$

Again, although many different situations are possible, two are common. When the random effects are uncorrelated, $\mathbf{\Phi}$ is a function of $q_r = r$ parameters

$$\mathbf{T} = \mathbf{T}(\boldsymbol{\tau}) = \mathrm{Diag}(\tau_1, \ldots, \tau_r)$$

with $[\mathbf{\Phi}]_{jj} = \tau_j^2$. If $\mathbf{\Phi}$ is a general symmetric matrix, then $\mathbf{T}$ is lower triangular,

$$\mathbf{T} = \mathbf{T}(\boldsymbol{\tau}) = \begin{pmatrix} t_{11} & \cdots & 0 \\ \vdots & \ddots & 0 \\ t_{r1} & \cdots & t_{rr} \end{pmatrix}$$

The elements of $\mathbf{T}$ correspond to those of $\boldsymbol{\tau}$ by the relationship $t_{jk} = \tau_L$, with

$$L(j, k) = k + j(j - 1)/2, \qquad j \geq k \tag{1.3}$$

The order of $\boldsymbol{\tau}$ when $\mathbf{\Phi}$ is symmetric is $q_\tau = r(r + 1)/2$ and $[\mathbf{\Phi}]_{jk} = \sum_{i=1}^{k} t_{ji} t_{ki}$.

The conditional distribution of $\mathbf{y}_i$ given $\mathbf{b}_i$ is

$$\mathbf{N}(\mathbf{f}_i(\boldsymbol{\beta} + \mathbf{Bb}_i + \boldsymbol{\Theta}\mathbf{z}_i, \mathbf{x}_i, \mathbf{z}_i), \boldsymbol{\Lambda}_i)$$

The density function is

$$f_{y|b}(\mathbf{y}_i \mid \mathbf{b}_i) = (2\pi)^{-\frac{n_i}{2}} |\boldsymbol{\Lambda}_i|^{-\frac{1}{2}} \exp\left[-\tfrac{1}{2}(\mathbf{y}_i - \mathbf{f}_i)'\boldsymbol{\Lambda}_i^{-1}(\mathbf{y}_i - \mathbf{f}_i)\right]$$

where $\mathbf{f}_i = \mathbf{f}_i(\boldsymbol{\beta}_i, \mathbf{x}_i, \mathbf{z}_i)$. The density function of $\mathbf{b}_i$ is

$$g(\mathbf{b}_i) = (2\pi)^{-\frac{r}{2}} |\mathbf{\Phi}|^{-\frac{1}{2}} \exp(-\tfrac{1}{2}\mathbf{b}_i'\mathbf{\Phi}^{-1}\mathbf{b}_i) \tag{1.4}$$

Therefore, the joint distribution of $\mathbf{y}_i$ and $\mathbf{b}_i$ is

$$\begin{aligned} f_{y,b}(\mathbf{y}_i, \mathbf{b}_i) &= f_{y|b}(\mathbf{y}_i \mid \mathbf{b}_i)g(\mathbf{b}_i) \\ &= K^* \cdot \exp\left[-\tfrac{1}{2}(\mathbf{y}_i - \mathbf{f}_i)'\boldsymbol{\Lambda}_i^{-1}(\mathbf{y}_i - \mathbf{f}_i)\right] \cdot \exp(-\tfrac{1}{2}\mathbf{b}_i'\mathbf{\Phi}^{-1}\mathbf{b}_i) \end{aligned}$$

where $K^* = (2\pi)^{-\frac{(r+n_i)}{2}} |\boldsymbol{\Lambda}_i|^{-\frac{1}{2}} |\mathbf{\Phi}|^{-\frac{1}{2}}$, and the unconditional distribution of $\mathbf{y}_i$ is

$$h_I(\mathbf{y}_i) = \int f_{y|b}(\mathbf{y}_i \mid \mathbf{b})g(\mathbf{b})d\mathbf{b} = \int f_{y,b}(\mathbf{y}_i, \mathbf{b})d\mathbf{b} \tag{1.5}$$

For a sample of N independent observations, $\mathbf{y} = (\mathbf{y}_1', \ldots, \mathbf{y}_N')'$, the likelihood function from the marginal distribution $h_I(\mathbf{y}_i)$ is $L_I(\boldsymbol{\beta}, \boldsymbol{\theta}, \boldsymbol{\tau}, \boldsymbol{\lambda} \mid \mathbf{y}) = \prod_{i=1}^{N} h_I(\mathbf{y}_i)$, with log-likelihood

$$\ln L_I(\boldsymbol{\beta}, \boldsymbol{\theta}, \boldsymbol{\tau}, \boldsymbol{\lambda} \mid \mathbf{y}) = \sum_{i=1}^{N} \ln h_I(\mathbf{y}_i) \tag{1.6}$$

In general, the integral in (1.5) cannot be computed explicitly. However, a satisfactory numerical approximation can be implemented with

Gauss-Hermite quadrature, after an appropriate change of variables. Let $\mathbf{T}^{-1}$ be the inverse of the square root matrix in (1.2) and set $\mathbf{u} = 2^{-\frac{1}{2}}\mathbf{T}^{-1}\mathbf{b}_i$ so that $\mathbf{b}_i = \sqrt{2}\mathbf{T}\mathbf{u}$, with Jacobian

$$\left|\frac{\partial \mathbf{b}_i}{\partial \mathbf{u}'}\right| = 2^{\frac{r}{2}}|\mathbf{\Phi}|^{\frac{1}{2}}$$

Then, in terms of the transformed variable, the distribution function in (1.5) becomes

$$h_I(\mathbf{y}_i) = K \int \exp\left(-\tfrac{1}{2}\mathbf{e}_i'\mathbf{\Lambda}_i^{-1}\mathbf{e}_i\right)\exp(-\mathbf{u}'\mathbf{u})d\mathbf{u} \qquad (1.7)$$

where $\mathbf{e}_i = \mathbf{y}_i - \mathbf{f}_i(\boldsymbol{\beta}_u, \mathbf{x}_i, \mathbf{z}_i)$,

$$\boldsymbol{\beta}_u = \boldsymbol{\beta} + \mathbf{B}(\sqrt{2}\mathbf{T}\mathbf{u}) + \mathbf{\Theta}\mathbf{z}_i \qquad (1.8)$$

and

$$K = K^* \left|\frac{\partial \mathbf{b}_i}{\partial \mathbf{u}'}\right| = \left(2^{n_i}\pi^{(n_i+r)}\right)^{-\frac{1}{2}}|\mathbf{\Lambda}_i|^{-\frac{1}{2}}$$

The function under the integral in (1.7) is of the form $g(\mathbf{z}) = \exp(-\mathbf{z}'\mathbf{z})f^*(\mathbf{z})$, where $f^*(\mathbf{z})$ is a scalar-valued function of $\mathbf{z} = (z_1, \ldots, z_r)'$. The Gauss-Hermite quadrature formula to approximate an r-dimensional integral of this type by a sum is, using $r = 3$ for example,

$$\int g(\mathbf{z})d\mathbf{z} = \int\int\int e^{-z_1^2}e^{-z_2^2}e^{-z_3^2}f^*(z_1, z_2, z_3)d\mathbf{z}$$

$$\simeq \sum_{g_1=1}^{G}\sum_{g_2=1}^{G}\sum_{g_3=1}^{G}(w_{g_1}w_{g_2}w_{g_3})f^*(u_{g_1}, u_{g_2}, u_{g_3})$$

where w_{g_k} and u_{g_k} are, respectively, the quadrature weights and abscissas for an approximation with G points. It can be shown (e.g., Krommer & Ueberhuber, 1994, sec. 4.2.6, or Stroud & Secrest, 1966, sec. 1) that the approximation

$$\int e^{-z^2}f^*(z)dz = \sum_{g=1}^{G}w_g f^*(u_g)$$

is exact if $f^*(z)$ is a polynomial of degree $2G - 1$. It can also be shown (Krommer & Ueberhuber, 1994, sec. 4) that the approximation converges to the true integral as the number of quadrature points increases

$$\lim_{G\to\infty}\sum_{g=1}^{G}w_g f^*(u_g) = \int e^{-z^2}f^*(z)dz$$

Let $W_G = w_{g_1} \cdots w_{g_r}$. The corresponding approximation to (1.7) is

$$h(\mathbf{y}_i) = K \sum_{g_1=1}^{G} \cdots \sum_{g_r=1}^{G} W_G \exp\left(-\tfrac{1}{2}\mathbf{e}_i' \mathbf{\Lambda}_i^{-1} \mathbf{e}_i\right) \qquad (1.9)$$
$$\simeq h_I(\mathbf{y}_i).$$

The residual vector $\mathbf{e}_i$ used in (1.9) is a function of $\boldsymbol{\beta}_u$ in (1.8) with abscissas, $u_{g_1}, \ldots, u_{g_r}$.

APPROXIMATING THE LIKELIHOOD FUNCTION

For marginal maximum likelihood estimation of the model, we define the log-likelihood as

$$\ln L(\boldsymbol{\beta}, \boldsymbol{\theta}, \boldsymbol{\tau}, \boldsymbol{\lambda} \mid \mathbf{Y}) = \sum_{i=1}^{N} \ln h(\mathbf{y}_i) \qquad (1.10)$$

with $h(\mathbf{y}_i)$ given in (1.9). For the Newton-Raphson method, the derivatives of the log-likelihood are needed. It is important that (1.10) be defined as the objective function rather than (1.6) with (1.7) because the gradient vectors of (1.6) and (1.10) are not the same. To see this, it is sufficient to show only the results pertaining to $\boldsymbol{\tau}$. As given in the Appendix, after the change of variables, the gradient of (1.6) with respect to $t_{jk} = [\mathbf{T}]_{jk}$ is

$$\frac{\partial \ln L_I(\boldsymbol{\beta}, \boldsymbol{\theta}, \boldsymbol{\tau}, \boldsymbol{\lambda} \mid \mathbf{Y})}{\partial t_{jk}} = \sum_{i=1}^{N} \frac{\partial \ln h_I(\mathbf{y}_i)}{\partial t_{jk}} \qquad (1.11)$$

where

$$\frac{\partial \ln h_I(\mathbf{y}_i)}{\partial t_{jk}} =$$
$$\frac{K}{h_I(\mathbf{y}_i)} \int \left[\mathbf{u}' \mathbf{T}^{-1} \dot{\mathbf{\Phi}}_{jk} (\mathbf{T}')^{-1} \mathbf{u} \right] \exp(-\tfrac{1}{2}\mathbf{u}'\mathbf{u}) m_i du - \frac{1}{2}\text{tr}\left[\mathbf{\Phi}^{-1} \dot{\mathbf{\Phi}}_{jk} \right] \qquad (1.12)$$

with $\mathbf{u} = (u_{g_1}, \ldots, u_{g_r})'$, $m_i = \exp(-\tfrac{1}{2}\mathbf{e}_i' \mathbf{\Lambda}_i^{-1} \mathbf{e}_i)$,

$$\dot{\mathbf{\Phi}}_{jk} = \frac{\partial \mathbf{\Phi}}{\partial t_{jk}} = \dot{\mathbf{T}}_{jk} \mathbf{T}' + \mathbf{T}\dot{\mathbf{T}}_{jk}'$$

and $\dot{\mathbf{T}}_{jk} = \frac{\partial \mathbf{T}}{\partial t_{jk}}$. As with the log-likelihood in (1.6), the integral in (1.12) must be computed numerically, again with Gauss quadrature

$$\frac{\partial \ln h_I(\mathbf{y}_i)}{\partial t_{rs}} \approx \frac{2K}{h(\mathbf{y}_i)} \sum_{g_1=1}^{G} \cdots \sum_{g_r=1}^{G} W_G m_i \left[\mathbf{u}' \mathbf{T}^{-1} \dot{\mathbf{T}}_{rs} \mathbf{u} \right] - \text{tr} \left[\mathbf{T}^{-1} \dot{\mathbf{T}}_{rs} \right]$$

(1.13)

On the other hand, if the log-likelihood is defined by (1.10) rather than (1.6), then the corresponding partial derivative is

$$\frac{\partial \ln L(\boldsymbol{\beta}, \boldsymbol{\theta}, \boldsymbol{\tau}, \boldsymbol{\lambda} \mid \mathbf{Y})}{\partial t_{jk}} = \sum_{i=1}^{N} \frac{\partial \ln h(\mathbf{y}_i)}{\partial t_{jk}}$$

(1.14)

where

$$\frac{\partial \ln h(\mathbf{y}_i)}{\partial t_{jk}} = \sqrt{2} \frac{K}{h(\mathbf{y}_i)} \sum_{g_1=1}^{G} \cdots \sum_{g_r=1}^{G} W_G m_i u_{g_k} [\mathbf{B}' \boldsymbol{\Delta}_i' \boldsymbol{\Lambda}_i^{-1} \mathbf{e}_i]_j$$

(1.15)

and

$$\boldsymbol{\Delta}_i = \frac{\partial \mathbf{f}_i(\boldsymbol{\beta}_u, \mathbf{x}_i)}{\partial \boldsymbol{\beta}_u'}$$

(1.16)

and where the subscript of the abscissa u_{g_k} is associated with one of the index variables, $g_k \in (g_1, \ldots, g_r)$. In practice, the difference in values obtained from (1.15) and (1.13) will depend on the model and the number of quadrature points.

In summary, because (1.10) with (1.9) is the actual way the function is implemented, it follows that the gradient vector must be calculated from (1.10), as is illustrated with the partial derivative of $\ln L(\boldsymbol{\beta}, \boldsymbol{\theta}, \boldsymbol{\tau}, \boldsymbol{\lambda} \mid \mathbf{Y})$ with respect to $\boldsymbol{\tau}$ in (1.14) and (1.15). Use of (1.11) with (1.13) does not give the proper maximum of the function computed in practice, so that gradient methods developed from it may either fail or at least perform suboptimally, and the asymptotic covariance matrix of the estimates would not be correct.

ESTIMATING THE MODEL

In this section, we present the steps needed to estimate the parameters of the model based on the Newton-Raphson algorithm, followed by estimation of the individual coefficients. The gradient vector for this problem is straightforward to compute. In contrast, elements of the Hessian matrix are not only complicated in form but are also relatively time-consuming to calculate. Consequently, we propose approximating the

Hessian numerically, using the explicit first derivatives (Dennis & Schnabel, 1983, sec. 5.6). Not only is this much simpler in terms of computer code, it is also faster than using the exact matrix of second derivatives. Results from the two methods typically agree to five or six significant digits, so estimates and standard errors agree well for most practical purposes.

Parameter Estimation

The gradient vector of (1.10) with respect to the parameters $\gamma = (\beta', \theta', \tau', \lambda')'$ is composed of the submatrices

$$\mathbf{g} = \mathbf{g}(\gamma) = \frac{\partial \ln L}{\partial \gamma} = (\mathbf{g}_\beta', \mathbf{g}_\theta', \mathbf{g}_\tau', \mathbf{g}_\lambda')'$$

The matrix of the Jacobian of the response function with respect to β_u is (1.16). Let the product of weights be $W_G = w_{g_1} \cdots w_{g_r}$. Define $m_i = \exp(-\frac{1}{2}\mathbf{e}_i'\Lambda_i^{-1}\mathbf{e}_i)$ and the following intermediate quantities

$$\dot{\Lambda}_j = \frac{\partial \Lambda}{\partial \lambda_j} \qquad \mathbf{a}_i = \Lambda_i^{-1}\mathbf{e}_i \qquad \mathbf{c}_i = \Delta_i'\mathbf{a}_i \qquad \dot{\mathbf{a}}_j = \dot{\Lambda}_j\mathbf{a}_i \qquad \alpha_i = \mathbf{B}'\mathbf{c}_i.$$

The first section in $\mathbf{g}$ is

$$\mathbf{g}_\beta = K \sum_{i=1}^{N} \frac{1}{h(\mathbf{y}_i)} \sum_{g_1=1}^{G} \cdots \sum_{g_r=1}^{G} W_G m_i \mathbf{c}_i$$

Elements of $\mathbf{g}_\theta$ corresponding to $[\Theta]_s$ are

$$\frac{\partial \ln L}{\partial [\Theta]_s} = K \sum_{i=1}^{N} \frac{1}{h(\mathbf{y}_i)} \sum_{g_1=1}^{G} \cdots \sum_{g_r=1}^{G} W_G m_i \mathbf{c}_i [\mathbf{z}_i]_s \qquad 1 \le s \le v$$

Next, using the identity $t_{jk} = \tau_L$, with L defined in (1.3), the second section is [cf. (1.14) and (1.15)]

$$[\mathbf{g}_\tau]_L = \sum_{i=1}^{N} \frac{1}{h(\mathbf{y}_i)} \frac{\partial h(\mathbf{y}_i)}{\partial t_{jk}} \qquad 1 \le k \le j \le r$$

where

$$\frac{\partial h(\mathbf{y}_i)}{\partial t_{jk}} = \sqrt{2}K \sum_{g_1=1}^{G} \cdots \sum_{g_r=1}^{G} W_G m_i u_{g_k} [\alpha_i]_j$$

Elements of the last section are

$$[\mathbf{g}_\lambda]_j = -\frac{1}{2} \sum_{i=1}^{N} \left[\mathrm{tr}\left(\Lambda^{-1}\dot{\Lambda}_j\right) - \frac{1}{S_i} \sum_{g_1=1}^{G} \cdots \sum_{g_r=1}^{G} W_G m_i \mathbf{a}_i'\dot{\mathbf{a}}_j \right] \qquad 1 \le j \le q_\lambda$$

where

$$S_i = \frac{h(\mathbf{y}_i)}{K} = \sum_{g_1=1}^{G} \cdots \sum_{g_r=1}^{G} W_G m_i \qquad (1.17)$$

The Hessian matrix has components

$$\mathbf{H}(\boldsymbol{\gamma}) = \frac{\partial^2 \ln L}{\partial \boldsymbol{\gamma} \partial \boldsymbol{\gamma}'} = \begin{pmatrix} \mathbf{H}_{\beta\beta} & & & \\ \mathbf{H}_{\theta\beta} & \mathbf{H}_{\theta\theta} & & \\ \mathbf{H}_{\tau\beta} & \mathbf{H}_{\tau\theta} & \mathbf{H}_{\tau\tau} & \\ \mathbf{H}_{\lambda\beta} & \mathbf{H}_{\lambda\theta} & \mathbf{H}_{\lambda\tau} & \mathbf{H}_{\lambda\lambda} \end{pmatrix}$$

The finite-difference approximation to $\mathbf{H}$ at a particular point $\boldsymbol{\gamma} = \boldsymbol{\gamma}_0$ using the explicit gradient vector is based on a related matrix, $\mathbf{H}^*(\boldsymbol{\gamma})$, that is built up column by column. Define the j-th column of this approximation as

$$[\mathbf{H}^*(\boldsymbol{\gamma})]_{.j} = \frac{\mathbf{g}(\boldsymbol{\gamma} + \eta_j \mathbf{v}_j) - \mathbf{g}(\boldsymbol{\gamma})}{\eta_j}$$

where $\mathbf{v}_j$ is a null vector except for a single value of unity in the j-th position, and η_j a small constant scaled in terms of parameter γ_j:

$$\eta_j = |\gamma_j| \cdot eps$$

where eps is a machine-dependent constant. For example, on microcomputers with 32-bit double precision word, one can set eps = $1.5x10^{-7}$. The matrix actually used in practice is computed from $\mathbf{H}^*(\boldsymbol{\gamma})$ in a way that insures the final matrix is symmetric

$$\mathbf{H}(\boldsymbol{\gamma}) = \frac{1}{2} \left[\mathbf{H}^*(\boldsymbol{\gamma}) + (\mathbf{H}^*(\boldsymbol{\gamma}))' \right]$$

The Newton-Raphson step from an estimate, $\boldsymbol{\gamma}^{(k)}$, of the parameters at the k-th iteration to the next point in the sequence, $\boldsymbol{\gamma}^{(k+1)}$, is the solution of the system of linear equations,

$$\mathbf{H}(\boldsymbol{\gamma}^{(k)})\boldsymbol{\delta} = -\mathbf{g}(\boldsymbol{\gamma}^{(k)})$$

with updated parameters

$$\boldsymbol{\gamma}^{(k+1)} = \boldsymbol{\gamma}^{(k)} + \omega \boldsymbol{\delta}$$

The step-halving coefficient, $\omega > 0$, is chosen to ensure that $\ln L(\boldsymbol{\gamma}^{(k+1)}) > \ln L(\boldsymbol{\gamma}^{(k)})$.

Define $\hat{\boldsymbol{\gamma}}$ as the maximizer of (1.10) that is hopefully obtained when the iterative process converges. In large samples, the standard error of estimate of element $[\hat{\boldsymbol{\gamma}}]_j$ is computed from the corresponding diagonal element of the approximate Hessian matrix in $se[\hat{\boldsymbol{\gamma}}]_j = [\mathbf{H}(\hat{\boldsymbol{\gamma}})^{-1}]_{jj}^{\frac{1}{2}}$. These are useful for constructing interval estimates and for testing hypotheses about the associated parameters.

Estimates of Individual Coefficients

After the maximum likelihood estimate $\hat{\gamma}$ has been computed, individual coefficients, $\mathbf{b}_i$, for the sum in (1.1) are estimated. These are the expected values of the conditional distribution, $f_{b|y}(\mathbf{b}_i| \mathbf{y}_i)$ where

$$\hat{\mathbf{b}}_i = \mathrm{E}(\mathbf{b}_i| \mathbf{y}_i) = \int \mathbf{b} f_{b|y}(\mathbf{b} | \mathbf{y}_i) d\mathbf{b}$$

From the joint density

$$f_{y,b}(\mathbf{y}_i, \mathbf{b}_i) = f_{y|b}(\mathbf{y}_i| \mathbf{b}_i)g(\mathbf{b}_i) = f_{b|y}(\mathbf{b}_i| \mathbf{y}_i)h_I(\mathbf{y}_i)$$

the conditional density function of $(\mathbf{b}_i| \mathbf{y}_i)$ can be written in terms of the density of $(\mathbf{y}_i| \mathbf{b}_i)$:

$$f_{b|y}(\mathbf{b}_i| \mathbf{y}_i) = \frac{f_{y|b}(\mathbf{y}_i| \mathbf{b}_i)g(\mathbf{b}_i)}{h_I(\mathbf{y}_i)}$$

This allows the expected value to be written as

$$\hat{\mathbf{b}}_i = h_I(\mathbf{y}_i)^{-1} \int \mathbf{b}\, f_{y|b}(\mathbf{y}_i| \mathbf{b})g(\mathbf{b})d\mathbf{b}$$

The integrals are again computed numerically by Gauss-Hermite quadrature. In this instance, the individual coefficients are the ratio

$$
\begin{aligned}
E(\mathbf{b}_i | \mathbf{y}_i) &= \frac{\sum_{g_1=1}^{G} \cdots \sum_{g_r=1}^{G} W_G m_i \mathbf{t}_u}{\sum_{g_1=1}^{G} \cdots \sum_{g_r=1}^{G} W_G m_i} \\
&= S_i^{-1} \sum_{g_1=1}^{G} \cdots \sum_{g_r=1}^{G} W_G m_i \mathbf{t}_u
\end{aligned}
\tag{1.18}
$$

where $\mathbf{t}_u = \sqrt{2}\mathbf{T}\mathbf{u}$, $\mathbf{u} = (u_{g_1}, \ldots, u_{g_r})'$, and S_i is in (1.17). Some estimators proposed in the literature use the posterior mode rather than the expected value for $\hat{\mathbf{b}}_i$. In this implementation, (1.18) is convenient computationally. Because of the assumption that $\mathbf{b}_i$ has a normal distribution, there should be little practical difference between the two.

EXAMPLES

Two examples are presented in this section.   The first includes a four-parameter function with one nonlinear coefficient.   The second example requires three nonlinear random effects plus three parameters for

TABLE 1.1

Bradway and McArdle's Longitudinal Study

	Occasions					
	1	2	3	4	5	6
Year	1931	1941	1956	1969	1984	1992
Median Age	4.3	14.1	29.9	42.9	57.7	66.2
N	74	74	74	56	53	54

Note. Subjects with at least one measurement on occasion 4 - 6.

a level-two covariate. Although these are not highly complex nonlinear functions compared to some examples that have been reported, the research questions are realistic and important in their own contexts. The examples show that this approach to nonlinear mixed-effects models is practical. Many interesting studies with a repeated measures component can be satisfactorily investigated using models that require only three or four random effects.

Data From a Longitudinal Study

One of the longest running longitudinal studies ever conducted was begun by Bradway (Bradway & Thompson, 1962) and extended by McArdle (McArdle & Hamagami, 1996). In 1931, 138 children were initially administered the Stanford-Binet. They were retested as many as five more occasions up to 1992 on both the Stanford-Binet and the Wechsler scales. A summary of the study is shown in Table 1.1. Our analysis is based on $N = 74$ subjects who had at least one score on occasions 4, 5, or 6. Thirty of these subjects were measured on all six occasions. The other 44 cases had one or two missing scores in six different missing data patterns. The response variable is a weighted sum of nonverbal items from both the Stanford-Binet and Wechsler scales, centered at the mean of the 1931 sample. Exact age was recorded at each occasion, so values on the predictor differ for each person. Records for a random 25% subsample are shown in Fig. 1.1. No level-two covariates are included in this analysis.

As shown in Fig. 1.1, there was an initial rapid growth phase until the early teenage years, at which point, performance leveled off. In late adulthood, some of the subjects had scores that gradually increased, whereas others declined. Although the overall pattern was similar for the sample as a whole, individual differences, as always, were a notable feature of the data.

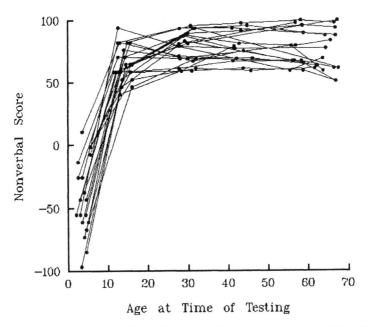

FIG. 1.1. Bradway and McArdle data: Measures of nonverbal intelligence over the lifespan on 25% of random sample $N = 74$.

Many models can be considered for this situation. Functions with an asymptote seem inappropriate because the pattern of change was not uniformly increasing or decreasing. It is of interest to formally test whether overall improvement versus overall decline characterized the population. In an effort to simply describe the data, various two phase models, corresponding to the preadult and adult phases, were tried. A satisfactory representation was possible with a segmented polynomial (e.g., Cudeck & Klebe, in press; Morrell, Pearson, Carter, & Brant, 1995; Seber & Wild, 1989, chap. 9)

$$[\mathbf{f}_i(\boldsymbol{\beta}, \mathbf{x})]_j = \begin{cases} \beta_0 + \beta_1 x_j + \beta_2(x_j - \beta_3)^2 & x_j \le \beta_3 \\ \beta_0 + \beta_1 x_j & x_j > \beta_3 \end{cases}$$

using x_j is age at the j-th occasion of measurement. Two parameters are especially interesting in this context. The *transition point* of the function is β_3. It is the value of X at which the polynomial of the first phase

TABLE 1.2

Maximum Likelihood Estimates

j	$\hat{\beta}_j$ (se)		b_0	b_1	b_2	b_3
				$\hat{\Phi}$		
0	81.1 (2.20)	b_0	76.4			
1	-.141 (0.05)	b_1	-.011	.026		
2	-.572 (0.04)	b_2	.090	.007	.032	
3	18.6 (0.60)	b_3	3.68	.099	.408	9.99

Note. $\hat{\sigma}^2 = 77.5$

changes to the linear component of the second. Developmentally, this would be interpreted as the age when a qualitative change in rate of cognitive growth occurs. The second interesting parameter is β_1, the slope of the linear phase. It can be used to test whether there is overall improvement or decline across the life span. If no random effects are specified for the transition point, then the model can be fit by standard maximum likelihood for linear models. Because there is variability in the age at which the transition occurs, however, it is more reasonable to allow random effects on β_3 as well as on the other parameters. Individual coefficients are therefore $\beta_i = \beta + \mathbf{b}_i$ with $p = r = 4$. We take Φ to be symmetric, but specify homogeneous error variances as $\Lambda_i = \sigma^2 \mathbf{I}$.

The model was fit with $G = 16$ quadrature points. Maximum likelihood estimates are shown in Table 1.2. The fitted mean response, $\mathbf{f}(\hat{\beta}, x)$, is shown in Fig. 1.2. The estimated transition point for the population was $\hat{\beta}_3 = 18.6$ years, with $se(\hat{\beta}_3) = 0.60$. After that age there was on average a gradual decline in performance of approximately .14 points per year ($\hat{\beta}_1 = -.141$, $se(\hat{\beta}_1) = .05$). Although the trend decreased overall, a few individuals actually exhibited increases, whereas for others, the response was essentially constant into old age. Figure 1.3 shows fitted functions for a few selected cases. The two individuals in Fig. 1.3(a) had large differences in intercept, β_{i0} (70.8 vs. 91.9); those in Fig. 1.3(b) had large differences in slope, β_{i1} (-.32 vs. .04); those in Fig. 1.3(c) had large differences in transition age, β_{i3} (14.1 vs. 23.6).

In their analyses, McArdle and Hamagami (1996) examined several different models. Their best fitting model was a latent curve structure. It also showed a slight decline in nonverbal intelligence, but only in the later adult years. In addition to differences in the statistical model, McArdle and Hamagami used a larger sample of $N = 111$. Because in our simple two-phase model, the linear component was of special interest, and because we lacked information regarding the dropout mechanism that resulted

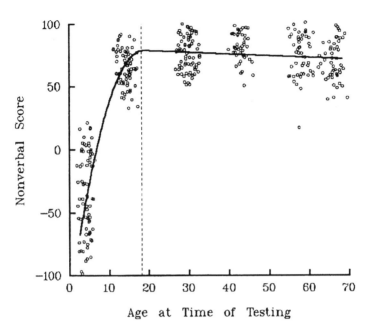

FIG. 1.2. Bradway-McArdle data: Sample data with fitted function.

in a smaller sample at the later measurement occasions, it was thought reasonable to require that subjects have at least one score in the last three periods. In light of this consideration, and in light of the interpretable and stable results, we believe these results are substantively reasonable, at least as a preliminary description.

A Verbal Learning Experiment

Smith and Klebe (1997) conducted an experiment in which $N = 143$ college students studied a list of 15 words during 10 trial periods. The number of items correctly recalled in each trial was recorded. In addition to the free recall experiment, a measure of verbal achievement, z_i, was obtained as a covariate. A 10% random sample of the data are shown in Fig. 1.4.

A three-parameter logistic response function seems appropriate for these data

$$[\mathbf{f}_i(\boldsymbol{\beta}_i, \mathbf{x}_i, \mathbf{z}_i)]_j = \frac{\beta_{i0}^*}{1 + \exp(\beta_{i1} - \beta_{i2}x_j)} \tag{1.19}$$

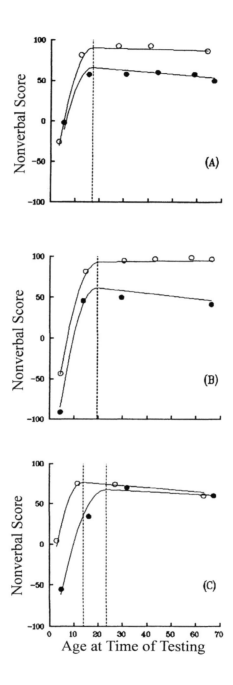

FIG. 1.3. Bradway-McArdle data: Fitted individual functions with (a) large differences in intercept, (b) large differences in slope, and (c) large differences in transition age.

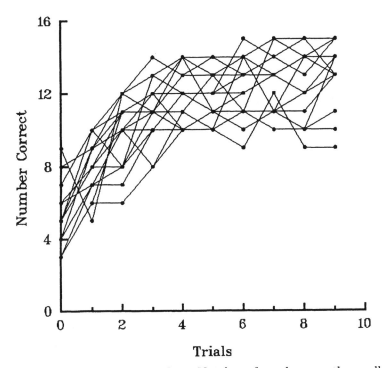

FIG. 1.4. Smith and Klebe's data: Number of words correctly recalled by 10% of sample, $N = 142$.

where $\mathbf{x} = (0, \ldots, 9)'$ are the trials. The maximum number correct in this experiment is 15, which is the upper bound for the asymptote, β_{i0}^*. If β_{i0}^* is assumed to have a normal distribution with mean β_0, then there may be cases for which values of $\beta_{i0}^* > 15$ occurs. It is more appropriate to assume that individual asymptotes have a skewed distribution and that coefficients are restricted to be $\beta_{i0}^* \leq 15$. This is handled by a transformation of the numerator of (1.19) in which

$$\beta_{i0}^*(\beta_{i0}) = \frac{15}{1 + e^{-\beta_{i0}}}$$

where the basic parameter β_{i0} together with β_{i1} and β_{i2} have a joint normal distribution, but $\beta_{i0}^*(\beta_{i0})$ is bounded above.

Two versions of this model were examined. In Model 1, individual coefficients are simply the sum

$$M_1 : \boldsymbol{\beta}_i = \boldsymbol{\beta} + \mathbf{b}_i$$

with $\mathbf{b}_i \sim \mathbf{N}(\mathbf{0}, \boldsymbol{\Phi})$. Here, $E(\boldsymbol{\beta}_i) = \boldsymbol{\beta}$. In Model 2, the covariate, z_i, is included at the second level to investigate the effectiveness of verbal achievement as a predictor of verbal learning. Let $\boldsymbol{\theta} = (\theta_0, \theta_1, \theta_2)'$ so that

$$M_2 : \boldsymbol{\beta}_i = \boldsymbol{\beta} + \mathbf{b}_i + \boldsymbol{\theta} z_i$$

with $E(\boldsymbol{\beta}_i) = \boldsymbol{\beta} + \boldsymbol{\theta}\mu_z$. It should be noted that (1.19) has only nonlinear parameters, and that the first is essentially a transformation of the distribution of β_{i0}. The performance of model M_2 against M_1 gives a means of assessing the improvement gained by adding the covariate. In each model, the residual structure was $\boldsymbol{\Lambda}_i = \sigma^2 \mathbf{I}$.

The number of quadrature points was $G = 14$. For the two models, the value of (1.10) was $\ln L_1 = -2441.9$ and $\ln L_2 = -2426.7$, with a difference of $-2(\ln L_1 - \ln L_2) = 30.4$ on degrees of freedom difference, $df = 3$. At least nominally, the covariate improves performance. Maximum likelihood estimates for M_2 are shown in Table 1.3. The largest value of the estimated asymptotes in the sample is $\hat{\beta}_{i0}^* = 14.7$. The fitted mean curve using $E(\mathbf{f}_i(\hat{\boldsymbol{\beta}}_i, \mathbf{x}_i, z_i))$ where

$$\mathbf{f}_i(\hat{\boldsymbol{\beta}}_i, \mathbf{x}_i, z_i) = 15 \left[\left\{ 1 + \exp(-\hat{\beta}_{i0}) \right\} \left\{ 1 + \exp(\hat{\beta}_{i1} - \hat{\beta}_{i2}\mathbf{x}) \right\} \right]^{-1}$$

is shown in Fig. 1.5. In the second-order model, the covariate Z is most effective as a predictor of β_{i0}^*, based on $\hat{\theta}_0 = .044$, $se(\hat{\theta}_0) = .006$. The covariate also predicts β_{i2}, but only weakly so, $\hat{\theta}_2 = .009$, $se(\hat{\theta}_2) = .002$. This suggests that verbal achievement is moderately successful in predicting final learning, β_{i0}^*, and perhaps also rate of learning, β_{i2}. The weight for β_{i1} on Z, θ_1, could well be zero.

DISCUSSION

Accuracy of the estimates, as well as amount of computer time, are a function of the number of quadrature points. In deciding on an appropriate number for G, a few different values can be tried and results based on the smallest number associated with stable estimates can be reported. For example, Model 2 of Example 2 had $N = 142$, $r = 3$ random effects, $v = 1$ covariates, with a total of 13 parameters. Table 1.4 shows values of $\ln L(\gamma)$, $\hat{\boldsymbol{\beta}}$, and the required computer time per Newton-Raphson iteration for $G = 8, 10, 12, 14$. The estimates for $G = 12$ and $G = 14$ are similar, at least for two significant digits. Computer time essentially doubles for each

TABLE 1.3

Maximum Likelihood Estimates on Recall Data

| j | $\hat{\beta}_j$ (se) | $\hat{\theta}_j$ (se) | | $\hat{\Phi}$ (se) | | |
				b_0	b_1	b_2
0	-2.46 (.62)	.044 (.006)	b_0	.699 (.08)		
1	1.33 (.54)	-.010 (.005)	b_1	-.102 (.04)	.186 (.03)	
2	-.208 (.18)	.009 (.002)	b_2	.048 (.02)	-.028 (.01)	.124 (.02)

Note. $\hat{\sigma}^2 = 1.07$ (.06)

TABLE 1.4

Computer Time and Accuracy of Estimates for Different Values of G

G	Time	$\ln L$	$\hat{\beta}_0$	$\hat{\beta}_1$	$\hat{\beta}_2$
8	4	-2433.9	-2.32	1.57	-.218
10	8	-2427.8	-2.42	1.48	-.213
12	14	-2426.1	-2.45	1.40	-.209
14	23	-2428.3	-2.46	1.33	-.208

increase in G to an elapsed time of 23 seconds per iteration for $G = 14$. The method can be demanding when G is large, but not prohibitively so.

This work demonstrates that marginal maximum likelihood estimation of the nonlinear mixed-effects models can be obtained by a direct implementation of the Gauss-Hermite formulas for numerical integration. The primary reason for this approach is to ensure that the likelihood function and its derivatives are mutually consistent. For any gradient-based method of estimation, it is essential that this be so. In addition to this important technical consideration, the method is practical. Many nonlinear models for repeated measures exist with a small number of random effects. With four random effects or fewer, this approach is feasible.

Much of the literature on multilevel models makes a qualitative distinction between linear and nonlinear models. Estimation in the former case is straightforward and software is widely available to fit models based on maximum likelihood assuming a normal distribution for both the residuals and random effects. Maximum likelihood estimation of nonlinear models is a much more difficult enterprise, in terms of derivation, computer code, and computational effort in the estimation step. Consequently, a good deal of statistical development has been concerned with approximations

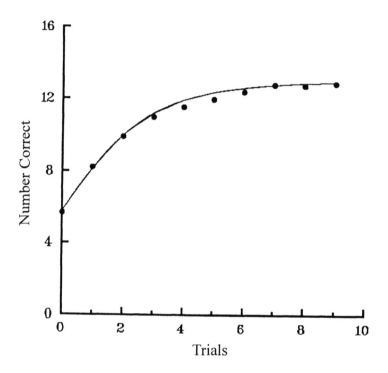

FIG. 1.5. Smith and Klebe's data: Sample means and fitted function.

to the general nonlinear model that result in estimators that are more manageable. These approximations work well in a variety of situations (e.g., Pinheiro & Bates, 1995; Vonesh & Carter, 1992). Nonetheless, it is more atisfying, at least theoretically, to approach mixed-effects models from a unified perspective that allows both linear and nonlinear response functions to be treated in the same way. To some extent, there is a corresponding distinction in the literature on linear versus nonlinear regression. A valuable feature of modern approaches to regression is the unified treatment of both classes of response function under maximum likelihood. This has illuminated a variety of subtle and not so subtle features of the model that would not be as clearly rendered from a compartmentalized approach. It is likely that further study of the mixed-effects model from a single perspective would be similarly advantageous.

Although the approach described here is successful as a method for estimating the nonlinear mixed-effects model, a sobering consideration in its use is the computational burden. Research is being conducted to investigate ways of reducing the computations. It appears that substantial improvements can be made in the way the numerical integration is handled. This will speed up the estimation process appreciably.

ACKNOWLEDGMENTS

We are grateful to Jack McArdle for allowing use of the Bradway–McArdle Longitudinal data, and to Kelli Klebe for data from her free recall experiment.   This research was partialy supported by NIMH grant MH5-4576.

REFERENCES

Bock, R. D. (1989). Measurement of human variation: A two–stage model. In R. D. Bock (Ed.), *Multilevel analysis of educational data* (pp. 319–342). San Diego: Academic Press.

Bradway, K., & Thompson, C. (1962). Intelligence at adulthood: A twenty-five year followup. *Journal of Educational Psychology, 53,* 1–14.

Bryk, A. S., & Raudenbush, S. W. (1992). *Hierarchical linear models: Applications and data analysis methods.* Newbury Park, CA: Sage.

Burden, R. L., & Faires, J. D. (1993). *Numerical analysis* (4th ed.). Boston: PWS-Kent.

Chambers, J. M., & Hastie, T. J. (Eds.) (1992). *Statistical Models in S.* Pacific Grove, CA: Wadsworth & Brooks/Cole.

Cudeck, R., & Kleve, K. J. (in press). Multiphase mixed-effects models for repeated measures data. *Psychological Methods.*

Davidian, M., & Giltinan, D. M. (1993). Some general estimation methods for nonlinear mixed-effects models. *Journal of Biopharmaceutical Statistics, 3,* 23–55.

Davidian, M., & Giltinan, D. M. (1995). *Nonlinear models for repeated measurement data.* London: Chapman & Hall.

Dennis, J. E., & Schnabel, R. B. (1983). *Numerical methods for unconstrained optimization and nonlinear equations.* Englewood Cliffs, NJ: Prentice-Hall.

Du Toit, S. H. C. (1993). *Analysis of multilevel models. Part 1: Theoretical aspects.* Pretoria, South Africa: Human Sciences Research Council.

Goldstein, H. (1987). *Multilevel models in educational and social research.* London: Griffin.

Kreft, I. G., De Leeuw, J., & Van der Leeden, R. (1994). Review of five multilevel analysis programs: BMDP-5V, GENMOD, HLM, ML3, VARCL. *American Statistician, 48*, 324–335.

Krommer, A. R., & Ueberhuber, C. W. (1994). *Numerical integration on advanced computer systems.* New York: Springer-Verlag.

Lindstrom, M. J., & Bates, D. M. (1990). Nonlinear mixed effects models for repeated measures data. *Biometrics, 46*, 673–687.

McArdle, J. J., & Hamagami, F. (1996). Multilevel models from a multiple group structural equation perspective. In G. A. Marcoulides & R. E. Schumacker (Eds.), *Advanced structural equation modeling* (pp. 89–124). Mahwah, NJ: Lawrence Erlbaum Associates.

Morrell, C. H., Pearson, J. D., Carter, H. B., & Brant, L. J. (1995). Estimating unknown transition times using a piecewise nonlinear mixed-effects model in men with prostrate cancer. *Journal of the American Statistician Association, 90*, 45–53.

Pinheiro, J. C., & Bates, D. M. (1995). Approximations to the loglikelihood function in the nonlinear mixed effects model. *Journal of Computational and graphical Statistics, 4*, 12–35.

Roe, D. J. (1997). Comparison of population pharmacokinetic modeling methods using simulated data: Results from the population modeling workgroup (with discussion). *Statistics in Medicine, 16*, 1241–1262.

Seber, G. A. F., & Wild, C. J. (1989). *Nonlinear regression.* NY: Wiley.

Smith, J. L., & Klebe, K. J. June, 1997, *Design issues in learning experiments: A comparison between repeated measures ANOVA and a latent growth curve.* Paper presented at the Annual Meeting of the Psychometric Society, Gatlinberg, TN.

Stroud, A. H., & Secrest, D. (1966). *Gaussian quadrature formulas.* New Jersey: Prentice–Hall.

Vonesh, E. F. (1992). Non-linear models for the analysis of longitudinal data. *Statistics in Medicine, 11*, 1929–1954.

Vonesh, E. F., & Carter, R. L. (1992). Mixed-effects nonlinear regression for unbalanced repeated measures. *Biometrics, 48*, 1–17.

Vonesh, E. F., & Chinchilli, V. M. (1997). *Linear and nonlinear models for the analysis of repeated measurements.* New York: Marcel Dekker.

Walker, S. (1996). An EM algorithm for nonlinear random effects models. *Biometrics, 52*, 934–944.

APPENDIX

In this section, we derive $\frac{\partial}{\partial \varphi_L} \ln h_I(\mathbf{y}_i)$, which is needed in the partial derivatives of (1.6) with respect to elements of τ. Define φ as the nonduplicated elements of Φ, with $\dot{\Phi}_L = \frac{\partial}{\partial \varphi_L}\Phi$. It is convenient to first find $\frac{\partial}{\partial \varphi_L} \ln h_I(\mathbf{y}_i)$. From (1.4),

$$\ln(g(\mathbf{b}_i)) = -\tfrac{r}{2}\ln(2\pi) - \tfrac{1}{2}\ln|\Phi| - \tfrac{1}{2}\mathbf{b}_i'\Phi^{-1}\mathbf{b}_i$$

Du Toit (1993, sec. 4) showed that

$$\frac{\partial \ln g(\mathbf{b}_i)}{\partial \varphi_L} = \tfrac{1}{2}\mathbf{b}_i'\Phi^{-1}\dot{\Phi}_L\Phi^{-1}\mathbf{b}_i - \tfrac{1}{2}tr\left[\Phi^{-1}\dot{\Phi}_L\right] \tag{A1}$$

From (1.5) (cf. Du Toit, 1993, sec. 5),

$$\frac{\partial \ln h_I(\mathbf{y}_i)}{\partial \varphi_L} = \frac{1}{h_I(\mathbf{y}_i)} \int f_{y|b}(\mathbf{y}_i \mid \mathbf{b})\left[\frac{\partial}{\partial \varphi_L}g(\mathbf{b})\right]d\mathbf{b} \tag{A2}$$

Because $\frac{\partial}{\partial \varphi_L}\ln g(\mathbf{b}_i) = g(\mathbf{b}_i)^{-1}\frac{\partial}{\partial \varphi_L}g(\mathbf{b}_i)$, it follows that

$$\frac{\partial g(\mathbf{b}_i)}{\partial \varphi_L} = g(\mathbf{b}_i)\frac{\partial \ln g(\mathbf{b}_i)}{\partial \varphi_L} \tag{A3}$$

Substituting (A3) into (A2),

$$\frac{\partial \ln h_I(\mathbf{y}_i)}{\partial \varphi_L} = \frac{1}{h_I(\mathbf{y}_i)} \int \frac{\partial \ln g(\mathbf{b})}{\partial \varphi_L} f_{y|b}(\mathbf{y}_i \mid \mathbf{b})g(\mathbf{b})d\mathbf{b} \tag{A4}$$

and (A1) into (A4) gives

$$\begin{aligned}\frac{\partial \ln h_I(\mathbf{y}_i)}{\partial \varphi_L} &= \frac{1}{2h_I(\mathbf{y}_i)} \int \left(\mathbf{b}'\Phi^{-1}\dot{\Phi}_L\Phi^{-1}\mathbf{b}\right) f_{y|b}(\mathbf{y}_i \mid \mathbf{b})g(\mathbf{b})d\mathbf{b}\\ &\quad -\frac{1}{2}tr\left[\Phi^{-1}\dot{\Phi}_L\right]\end{aligned} \tag{A5}$$

Let $K^* = (2\pi)^{-\frac{(r+n_i)}{2}}|\Lambda_i|^{-\frac{1}{2}}|\Phi|^{-\frac{1}{2}}$. Substituting $f_{y|b}(\mathbf{y}_i \mid \mathbf{b}_i)$ and $g(\mathbf{b}_i)$ into (A6) gives

$$\begin{aligned}\frac{\partial \ln h_I(\mathbf{y}_i)}{\partial \varphi_L} &= \frac{K^*}{2h_I(\mathbf{y}_i)} \int \left(\mathbf{b}'\Phi^{-1}\dot{\Phi}_L\Phi^{-1}\mathbf{b}\right)\exp\left(-\tfrac{1}{2}\mathbf{e}_i'\Lambda_i^{-1}\mathbf{e}_i\right)\\ &\quad \exp\left(-\tfrac{1}{2}\mathbf{b}'\Phi^{-1}\mathbf{b}\right)d\mathbf{b} - \frac{1}{2}tr\left[\Phi^{-1}\dot{\Phi}_L\right]\end{aligned} \tag{A6}$$

With the change of variables to $\mathbf{u} = \frac{1}{\sqrt{2}}\mathbf{T}^{-1}\mathbf{b}_i$, (A6) becomes

$$\frac{\partial \ln h_I(\mathbf{y}_i)}{\partial \varphi_L} = \frac{K}{2h_I(\mathbf{y}_i)} \int \left[(\sqrt{2}\mathbf{Tu})' \mathbf{\Phi}^{-1}\dot{\mathbf{\Phi}}_L \mathbf{\Phi}^{-1}(\sqrt{2}\mathbf{Tu}) \right] \exp(-\tfrac{1}{2}\mathbf{u}'\mathbf{u})$$
$$m_i d\mathbf{u} - \frac{1}{2}\mathrm{tr}\left[\mathbf{\Phi}^{-1}\dot{\mathbf{\Phi}}_L \right]$$

with $K = 2^{\frac{r}{2}}|\mathbf{\Phi}|^{\frac{1}{2}} K^* = (2\pi)^{-\frac{n_i}{2}} |\mathbf{\Lambda}_i|^{-\frac{1}{2}}$ and $m_i = \exp(-\tfrac{1}{2}\mathbf{e}_i'\mathbf{\Lambda}_i^{-1}\mathbf{e}_i)$. Because $\mathbf{\Phi}$ depends on $\mathbf{T}$ in (1.2), substitution gives

$$\frac{\partial \ln h_I(\mathbf{y}_i)}{\partial \varphi_L} = \frac{K}{h_I(\mathbf{y}_i)} \int \left[\mathbf{u}'\mathbf{T}^{-1}\dot{\mathbf{\Phi}}_L (\mathbf{T}')^{-1}\mathbf{u} \right] \exp(-\tfrac{1}{2}\mathbf{u}'\mathbf{u}) m_i d\mathbf{u}$$
$$- \frac{1}{2}\mathrm{tr}\left[\mathbf{\Phi}^{-1}\dot{\mathbf{\Phi}}_L \right] \tag{A7}$$

Gauss-Hermite quadrature is used to approximate (A7) as follows:

$$\frac{\partial \ln h_I(\mathbf{y}_i)}{\partial \varphi_L} \approx \frac{K}{h(\mathbf{y}_i)} \sum_{g1=1}^{G} \cdots \sum_{gr=1}^{G} W_G m_i \left[\mathbf{u}'\mathbf{T}^{-1}\dot{\mathbf{\Phi}}_L (\mathbf{T}')^{-1}\mathbf{u} \right] - \frac{1}{2}\mathrm{tr}\left[\mathbf{\Phi}^{-1}\dot{\mathbf{\Phi}}_L \right]$$

where $W_G = w_{g1} \cdots w_{gr}$. Again, substituting for $\mathbf{\Phi}$ from (1.2) and noting that $\dot{\mathbf{\Phi}}_{jk} = \dot{\mathbf{T}}_{jk}\mathbf{T}' + \mathbf{T}\dot{\mathbf{T}}'_{jk}$ gives finally

$$\frac{\partial \ln h_I(\mathbf{y}_i)}{\partial t_{jk}} \approx \frac{2K}{h(\mathbf{y}_i)} \sum_{g1=1}^{G} \cdots \sum_{gr=1}^{G} W_G m_i \left[\mathbf{u}'\mathbf{T}^{-1}\dot{\mathbf{T}}_{jk}\mathbf{u} \right] - \mathrm{tr}\left[\mathbf{T}^{-1}\dot{\mathbf{T}}_{jk} \right]$$

2

Sensitivity Analysis for Hierarchical Models: Downweighting and Identifying Extreme Cases Using the t Distribution

Michael Seltzer
Kilchan Choi
University of California, Los Angeles

A key facet of data analysis entails checking the adequacy of models. It is important to learn whether one's results are being strongly influenced by one or two cases, whether one's results might be especially sensitive to certain modeling choices and assumptions (e.g., choice of link function, distributional assumptions regarding random effects), and whether one's model fails to capture important features of the data (e.g., nonlinear relationships between key predictors and the outcome of interest).

Normality assumptions are commonly employed in hierarchical modeling settings. For example, in the case of two-level hierarchical models (HMs) for continuous outcomes, level-1 (within–cluster) error terms are typically assumed to be normally distributed. At level 2, cluster effects (i.e., random effects) are generally assumed to be normally distributed as well. In the case of HMs with more than two levels, normality assumptions are also typically employed for random effects specified at higher levels of the hierarchy (e.g., level 3).

It is widely known that in fitting models under normality assumptions, parameter estimates are potentially vulnerable to extreme cases (e.g., Mosteller & Tukey, 1977). This problem has stimulated the development and use of an array of robust regression techniques.

In the case of HMs, some attention has been given to the sensitivity of fixed effects estimates to extreme level–2 units. Consider, for example, a multisite evaluation study in which treatment and control conditions have been implemented in each of a series of sites. Treating treatment/control group contrasts as outcomes in a level–2 (between–site) model, we might be especially interested in modeling differences in the magnitude of treatment/control group contrasts as a function of program implementation. A problem, however, is that under normality assumptions at level 2, a site at which a program was unusually successful or unsuccessful could strongly impact the resulting estimate of the implementation fixed effect (e.g., Seltzer, Wong, & Bryk, 1996). As a second example, consider a growth modeling study in which children's rates of change with respect to a cognitive skill of interest are, in a level–2 (between–child) model, modeled as a function of whether or not children have attended preschool. Under normality assumptions in the level–2 model, one or two children with unusually slow or rapid rates of change could strongly impact the estimation of the fixed effect capturing the relationship between preschool and rate of change.

To address problems of this kind, a number of researchers have presented strategies for conducting sensitivity analyses under t level–2 assumptions (e.g., Carlin, 1992; Seltzer, 1993). These strategies are based on the scale mixture of normals representation of the t (e.g., Dempster, Laird, & Rubin, 1980; Lange, Little, & Taylor, 1989; West, 1984). Such analyses can be carried out fairly readily using Markov Chain Monte Carlo (MCMC) techniques (discussed later). When we fix the degrees of freedom parameter of the t distribution (e.g., ν_2) at a small value – that is, when heavy tails are assumed at level 2 – extreme level–2 units are more easily accommodated than under normality assumptions. The net effect is that extreme level–2 units will be downweighted in our analyses. Analogous to robust regression techniques, ν_2 serves as a tuning parameter. As the value at which we fix ν_2 decreases, the extent to which level–2 outliers will be downweighted increases.

Careful application of HMs also requires that we attend to extreme level–1 units. In a multisite evaluation study, for example, a person in the treatment group at a particular site whose outcome score is very extreme in relation to the other individuals in that particular group would be an example of a level–1 outlier. In the context of growth modeling applications, a time-series observation for an individual that is unusually high or low given the overall trend in that person's data would be considered a level–1 outlier.

As Rachman–Moore and Wolfe (1984) pointed out, level–1 outliers can "sour" summaries of the data for a given cluster, which, in turn, can impact the estimation of fixed effects. Consider, for example, the simple one-way ANOVA setting with random effects, where individuals are nested in J different clusters. In this setting, OLS estimates of the means for the clusters $(\overline{Y}_j, \, j = 1, \ldots, J)$ are used in computing the generalized least squares estimate of the grand mean (e.g., Bryk & Raudenbush, 1992, chap 3). Clearly, an extreme score within a particular cluster can impact the estimate of the mean for that cluster, which, in turn, can impact the estimate of the grand mean. Analogously, level–1 outliers can impact estimates of treatment/control contrasts for particular sites, or estimates of growth rates for certain children, which, in turn, can result in misleading estimates of fixed effects of interest. To this end, Seltzer, Novak, Choi and Lim (in press) presented a strategy that entails employing the scale mixture of normals representation of the t at both levels 1 and 2 of HMs (cf. Spiegelhalter, Best, Gilks, & Inskip, 1996).

Through analyses of the data from a longitudinal study of change in toddlers' request behavior we will (a) further highlight possible problems connected with level–1 and level–2 outliers, (b) illustrate the value of conducting sensitivity analyses under t distributional assumptions at levels 1 and 2 of HMs, and (c) illustrate how such analyses can be carried out using the software package WinBUGS (Spiegelhalter, Thomas, & Best, 2000), which was developed by statisticians in the MRC Biostatistics Unit in Cambridge, England. BUGS is a near acronym for Bayesian inference using Gibbs sampling. There has been an explosion of interest in Gibbs sampling and other MCMC techniques in the last 10 years in the statistics community. This is due to the fact that MCMC provides a viable approach to statistical inference in many complex settings. WinBUGS is freely available via the web: www.mrc–bsu.cam.ac.uk/bugs/welcome.shtml, and can be used in a large array of modeling settings. Detailed descriptions of the BUGS code that we have written for this chapter can be downloaded from the following website: www.gseis.ucla.edu/faculty/pages/seltzer.html.

In the following section, we describe the data set that we use in our illustrative examples. This provides a backdrop for a brief discussion of our estimation approach. We then take the reader through a series of detailed analyses of the data.

GROWTH MODEL FOR THE REQUEST
BEHAVIOR DATA

When making requests of their caretakers, toddlers often employ hand and body movements that partially enact the actions that they want their caretakers to perform. Smiley, Greene, Seltzer, and Choi (2000) referred to such behavior as enactive gesturing. Smiley et al.'s interest in enactive gesturing stemmed from the potential insight that behaviors of this kind can provide regarding the nature of toddlers' knowledge of self and other.

In our illustrative examples, we utilize the data from a longitudinal study that explored changes in the use of enactive gestures by toddlers (Smiley et al., 2000). Our sample consisted of 9 children (5 females and 4 males) and their mothers. Mother–child dyads were recruited from a middle class, well–educated urban community through newspaper advertisements. The dyads were videotaped during the course of their normal daily activities; the length of each taping session ranged from 2.5 to 3.5 hours. All of the requests initiated by a child during a taping session were subsequently classified into various categories (e.g., initiated with enactive gestures; initiated without enactive gestures, etc.). The outcome variable of interest in our analyses was the percentage of requests that were initiated using enactive gestures (EG).

Figure 2.1 displays the observed EG trajectories for the children in our sample. The sample contained EG measures at ages 12, 16, 20, 24, and 30 months for children 1 through 8, and EG measures at ages 12, 16, and 24 months for child 9. At 12 months of age, EG percents ranged from 50.0 to 93.2, with a mean of 76.1 and a standard deviation of 15.0. The general pattern is that EG percents decreased in a fairly linear fashion during the 12– to 24–month age range, and then flattened out after 24 months. Note that EG values for the sample of children tended to converge at the 24–month time point. This lends some support to prior research that suggests that at approximately 24 months, children display behaviors that indicate they have begun to construct a notion of self as a psychological entity (see Smiley et al., 2000 for a review).

In addition to coding children's request behaviors, mother's responses following children's requests were categorized. Of particular interest in the Smiley et al. study were caretaker responses that referred to children's states, actions, or goals, termed "speech about the child" (SPAC). Percents of caretaker responses falling into this category were computed at each time point.

Over the 12– to 20–month age range, mothers differed substantially in terms of their levels of SPAC, but seemed to show little systematic increase or decrease in their SPAC values over time. One question of interest

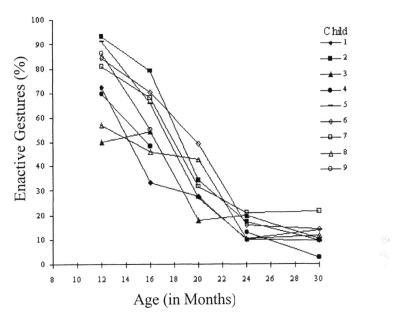

FIG. 2.1. The observed enactive gesturing (EG) trajectories for the 9 children in our sample.

investigated by Smiley et al. was the following: How do differences in levels of SPAC in this particular period relate to differences in rates of change in enactive gesturing (EG) over the 12– to 24–month age range. To examine this question, the SPAC values for each caretaker were averaged across the 12 to 20 month age span. The resulting variable (MSPAC) was employed as a level–2 (between–child) predictor in subsequent growth modeling analyses.

Smiley, Greene, Seltzer, and Choi (2000) modeled EG values across the entire age range (i.e., 12–30 months). In doing so, they employed a piecewise model for individual growth that captured rate of change during the 12 to 24 month age range, status at 24 months, and rate of change in the 24 to 30 month age range. For ease of exposition, we set aside the 30–month EG values and use a simpler model. Note that the following results are extremely similar to those reported by Smiley et al. (see Endnote 1). Note also that transforming EG values to the logit scale or using an arcsin transformation does not alter the pattern of results.

We now pose the following two–level growth model. We begin by specifying a level–1 (or within–child) model in which the series of EG values for each child in the 12– to 24–month age span is modeled as a linear function of age:

$$Y_{ti} = \pi_{0i} + \pi_{1i}(AGE_{ti} - 24) + \epsilon_{ti} \quad \epsilon_{ti} \sim N(0, \sigma^2), \quad (2.1)$$

where Y_{ti} represents the EG percent for child i ($i = 1, \ldots, I$) at measurement occasion t ($t = 1, \ldots, T_i$), AGE_{ti} represents the age in months of child i at measurement occasion t, and π_{1i} represents the rate of change for child i during the 12 to 24 month age range. As noted earlier, 24 months is viewed as a pivotal age in the development of autonomy. By virtue of centering AGE_{ti} around a value of 24, π_{0i} represents the EG status of child i at 24 months. Initially, we assume that the ϵ_{ti} (i.e., the level–1 residuals) are normally distributed with mean 0 and variance σ^2.

We then pose the following level–2 or between-child model:

$$\begin{aligned} \pi_{0i} &= \beta_{00} \\ \pi_{1i} &= \beta_{10} + \beta_{11}MSPAC_i + U_{1i} \quad U_{1i} \sim N(0, \tau_{11}). \end{aligned} \quad (2.2)$$

The parameter of primary interest in this model is β_{11}, which captures the relationship between level of mother's speech about the child (MSPAC) and rate of change in enactive gesturing. The parameter U_{1i} is a random effect that captures the deviation of the growth rate for child i from an expected value based on his or her caretaker's MSPAC value [i.e., $U_{1i} = \pi_{1i} - (\beta_{10} + \beta_{11}MSPAC_i)$]. We initially assume that the U_{1i} ($i = 1, \ldots, I$) are normally distributed with mean 0 and variance τ_{11}, where τ_{11} represents the variance in growth rates across children that remains after taking into account MSPAC.

In the first equation in the between-child model, β_{00} represents mean EG status at 24 months. Note that preliminary models that we fit to the data indicate that the variance in EG status at 24 months across children (τ_{00}) is extremely small, which is consistent with the pattern observed in Fig. 2.1. As such, as in Smiley et al. (2000), we do not include random effects for EG status in our level-2 model; that is, we constrain τ_{00} to a value of 0.

ESTIMATION AND INFERENCE

Fully Bayesian Analysis via MCMC

Typically, point estimates and standard errors for fixed and random effects in applications of HMs are based on the GLS and shrinkage estimation formulae outlined in such sources as Bryk and Raudenbush (1992). In these formulae, the variance components in HMs are assumed known. From a Bayesian perspective, these formulae correspond to the means and standard deviations of the conditional posterior distributions of the fixed and random effects given the data and given the variance components [e.g., $p(\beta_{11} \mid \mathbf{y}, \tau_{11}, \sigma^2)$]. Iterative techniques such as the EM algorithm or Fisher scoring are used to obtain ML estimates of the variance components, and these estimates are then substituted into the formulae for the fixed and random effects. Such an approach has been termed empirical Bayes (EB) (Bryk & Raudenbush, 1992). Thus, for example, the mean and standard deviation of the conditional posterior $p(\beta_{11} \mid \mathbf{y}, \tau_{11} = \hat{\tau}_{11}, \sigma^2 = \hat{\sigma}^2)$ would provide us with a point estimate and standard error for β_{11}.

In this approach, it can be seen that ML estimates of the variance components are essentially treated as the known true values of these parameters. This can be problematic in small-sample settings. That is, when the number of level–2 units in a sample is small, the EB approach can potentially result in underestimates of uncertainty (e.g., standard errors that are too small), and point estimates that may constitute poor summaries of the data (Draper, 1995; Rubin, 1981; Seltzer et al., 1996). (Regarding hypothesis tests and intervals for fixed effects in small-sample settings, note that the HLM program performs a correction that tends to provide appropriate rejection rates and levels of coverage provided that one's data are not too unbalanced.)

In contrast, the fully Bayesian (FB) approach entails basing inferences on the marginal posterior distributions of parameters of interest [e.g., $p(\beta_{11} \mid \mathbf{y})$]. This involves specifying prior distributions for the variance components, as well as all other parameters in one's model. To obtain the marginal posterior distribution of a parameter of interest, we integrate over all other parameters in the model. Thus, for example, $p(\beta_{11} \mid \mathbf{y})$ would provide a summary of the plausibility of different values for β_{11} given the data at hand and any available prior information. The mode, median, and mean of $p(\beta_{11} \mid \mathbf{y})$ would constitute various point estimates for β_{11}, and the .025 and .975 quantiles of this distribution would provide us with the Bayesian analogue of a 95% confidence interval.

One of the advantages of the FB approach is that it provides a general strategy for drawing inferences concerning a parameter of interest in a manner that takes into account the uncertainty connected with all other parameters in one's model. For example, in drawing inferences concerning β_{11}, integrating over τ_{11} and σ^2 along with all other unknowns in effect propagates the uncertainty concerning these parameters into $p(\beta_{11} \mid \mathbf{y})$ (Draper, 1995; Rubin, 1981; Seltzer et al., 1996; see, especially, Box & Tiao (1973, chap. 2) for a valuable discussion of this concept).

Calculating marginal posteriors of interest has heretofore been intractable in all but the simplest HM settings. Markov Chain Monte Carlo (MCMC) techniques, such as the Gibbs sampler, now make such an approach extremely viable (see, e.g., Carlin & Louis, 1996; Gelfand, Hills,

Racine-Poon, & Smith, 1990; Gelfand & Smith, 1990; Gelman, Carlin, Stern, & Rubin, 1995; Gilks, Richardson, & Spiegelhalter, 1996; Seltzer et al., 1996; Spiegelhalter, Thomas, et al., 1996b, 1996c; Spiegelhalter et al. 2000; Tanner, 1996; Tanner & Wong, 1987). As detailed in these references, MCMC techniques, in effect, provide a means of simulating marginal posteriors of interest in high-dimensional modeling settings. Examples of numerous applications of MCMC can be found in these references as well.

In addition to providing a viable strategy in settings in which the use of large-sample theory may be problematic, conducting fully Bayesian analyses via MCMC techniques places the data analyst in a position to capitalize on other important aspects of the Bayesian approach. In particular, the Bayesian approach encourages us to lay bare key assumptions in our models (e.g., distributional assumptions, specifications of priors) and to study the sensitivity of our results to sensible alternative assumptions. These ideas are extremely well articulated in the work of George Box (1979; 1980; see also Box & Tiao, 1973). MCMC greatly increases our capacity to put this important set of ideas into practice. In particular, we focus on the use of MCMC in conducting sensitivity analyses under t distributional assumptions.

Fitting HMs under t distributional assumptions

In our analyses, we employ a mixed modeling formulation. Thus, we collapse our level–1 and level–2 models (Equations 2.1 and 2.2) as follows:

$$Y_{ti} = \beta_{00} + [\beta_{10} + \beta_{11}MSPAC_i + U_{1i}](AGE_{ti} - 24) + \epsilon_{ti}. \quad (2.3)$$

As noted earlier, we initially assume normality at levels 1 and 2: $\epsilon_{ti} \sim N(0, \sigma^2)$ and $U_{1i} \sim N(0, \tau_{11})$. By virtue of assuming normality at level 1, we have

$$Y_{ti} \sim N(\mu_{ti}, \sigma^2), \quad (2.4)$$

where

$$\mu_{ti} = \beta_{00} + [\beta_{10} + \beta_{11}MSPAC_i + U_{1i}](AGE_{ti} - 24). \quad (2.5)$$

To implement the Gibbs sampler in settings in which we wish to employ t distributional assumptions at one or more levels of the algorithm, it is extremely convenient to work with the scale mixture of normals representation of the t. To help grasp the logic of this approach, recall that a t– distributed variate with mean 0, scale 1, and degrees of freedom ν can be expressed as: $z / \omega^{1/2}$, where z has a standard normal distribution $[z \sim N(0, 1)]$ and ω is a chi–squared distributed variate divided by its degrees of freedom ($\omega \sim \chi_\nu^2/\nu$). Note that the distribution χ_ν^2/ν

corresponds to a Gamma distribution with shape parameter $\nu/2$ and scale parameter $\nu/2$ [i.e., $Gamma(\nu/2, \nu/2)$] (see, e.g., Gelman et al., 1995, for details concerning the Gamma distribution).

Building on this logic, if we wish to assume that level–1 errors are t distributed with mean 0, scale σ^2, and ν_1 degrees of freedom, $N(\mu_{ti}, \sigma^2)$ in Equation 2.4 is now replaced by:

$$Y_{ti} \sim N(\mu_{ti}, \sigma^2/\omega_{ti}), \qquad (2.6)$$

where

$$\omega_{ti} \sim Gamma(\nu_1/2, \nu_1/2). \qquad (2.7)$$

We refer to the ω_{ti} as level–1 weight parameters. In Equation 2.6, note that as ω_{ti} decreases, the variance of Y_{ti} (i.e., σ^2/ω_{ti}) increases. As can be seen in the illustrative examples, those level–1 observations with small weights will, as in a Weighted Least Squares analysis, be downweighted.

A Gibbs sampling algorithm for models of this kind is detailed in Seltzer et al. (in press). Similar to the weights produced by robust regression techniques such as biweighting, the algorithm produces estimates of the ω_{ti}. Specifically, the algorithm is likely to generate small values for ω_{ti} when the distance of Y_{ti} from μ_{ti} is large.

This strategy can also be easily extended to level 2. Under the assumption that the random effects in our model are t distributed with ν_2 degrees of freedom, we would use the scale mixture of normals formulation as follows:

$$U_{1i} \sim N(0, \tau_{11}/q_i), \qquad (2.8)$$

where:

$$q_i \sim Gamma(\nu_2/2, \nu_2/2). \qquad (2.9)$$

In this formulation, the q_i constitute level–2 weight parameters. A small estimate (e.g., a small posterior mean) for q_i would signal a child whose rate of change is unusually slow or rapid.

If we wish to assume heavy tails at levels 1 or 2, we can fix the corresponding degrees of freedom parameter (ν_1; ν_2) at a small value (e.g., 4). If results obtained under heavy–tailed assumptions differ substantially from those obtained under normality, a useful strategy entails fixing degrees of freedom parameters along a grid of values (described later). We will also see that it is possible to treat ν_1 and ν_2 as parameters that are estimated based on the data at hand and on available prior information. This results in estimates for parameters of interest (e.g., fixed effects) that are analogous to robust adaptive estimates.

Implementation

Deriving the steps of Gibbs sampling algorithms can be a fairly complex task. Furthermore, implementing Gibbs sampling algorithms in languages such as Fortran can be extremely time–consuming. Fortunately, the software package WinBUGS, which we used to conduct the analyses presented in this chapter, provides a relatively easy means of implementing the Gibbs sampler in a wide range of modeling settings. In addition, the developers of WinBUGS have made available a comprehensive suite of programs for assessing convergence called <u>CODA</u> (Best, Cowles, & Vines, 1996). Like WinBUGS, CODA is freely available, and can be obtained via the BUGS website.

In endnote 3, we discuss the procedures that we used to assess convergence, and in endnote 4, we discuss the specification of priors for the variance components and fixed effects in our models. All analyses were run on a Pentium II 400mhz PC. To ensure high degrees of accuracy in simulating marginal posteriors of interest, all posterior means, standard deviations, and intervals reported are based on chains of 40,000 values generated by the Gibbs sampler (see endnote 3). For all analyses except those in which degrees of freedom parameters were treated as unknowns, less than 1 minute of CPU time was required to complete 40,000 iterations of the Gibbs sampler; when degrees of freedom parameters were treated as unknowns, approximately 4 minutes of CPU time were required.

ILLUSTRATIVE EXAMPLE 1

We first consider the growth model for the enactive gesturing data in which normality is assumed at levels 1 and 2 (N/N) (see Equation 2.3). Note that the HLM program (Bryk, Raudenbush, & Congdon, 1996) produces files termed residual files that can be used to obtain EB and Least Squares (LS) growth parameter estimates for the individuals in a sample (see endnote 2). In building HMs and checking their fit, we often find it useful to examine plots of the LS estimates produced by HLM versus level–2 predictor variables. Figure 2.2 displays the LS estimates of the rates of change for the 9 children in our sample versus MSPAC. As can be seen, as MSPAC increases, the LS estimates of the π_{1i} become more negative; that is, higher MSPAC values are associated with more rapid rates of decline in enactive gesturing.

We now use WinBUGS to fit the HM specified in Equation 2.3. Note that we have centered MSPAC around its grand mean. By virtue of this, β_{10} represents the mean rate of change in enactive gesturing during the 12– to 24–month age range. For β_{10}, we see in Table 2.1 that the mean of the

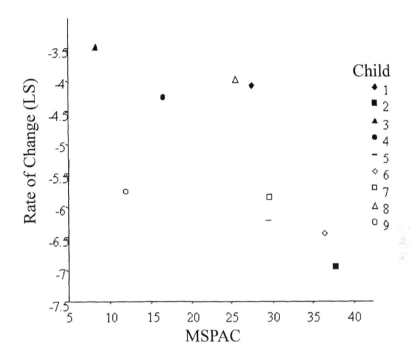

FIG. 2.2.   Illustrative Example 1: Least squares estimates of rates of change in enactive gesturing versus level of mother's speech about the child (see endnote 2).

resulting marginal posterior distribution is equal to a value of -5.23. One can interpret marginal posterior means just as one would interpret point estimates obtained via programs such as HLM. Thus, this result indicates that EG percents are, on average, declining at a rate of approximately $5\frac{1}{4}$ points per month during the 12– to 24–month age range. For β_{00} (the fixed effect representing mean status at 24 months), the marginal posterior mean takes on a value of 13.70 percentage points.

Turning to the results for β_{11} (the fixed effect capturing the relationship between MSPAC and rate of change), we see that the mean of the marginal posterior distribution takes on a value of -0.082, and that the lower and upper boundaries of the 95% interval based on this distribution take on values of -0.165 and -0.0003, respectively. It can also be seen that a value of 0 lies just above the upper boundary of the 95% interval. Note that approximately 2.5% of the marginal posterior for β_{11} lies above a value of

TABLE 2.1
Example 1: Posterior distributions for mean status at 24 months, mean rate, and the fixed effect for MSPAC (mother's speech about the child) under $N/N, t_4/N, N/t_4, t_4/t_4$, and t_3/t_4 level-1 / level-2 distributional assumptions.

	Mean	SD	95% Interval	Prob.> 0 [a]
Mean Status (2 yrs)				
$p(\beta_{00}\vert y)_{N/N}$	13.70	2.57	(8.68, 18.81)	
$p(\beta_{00}\vert y)_{t_4/N}$	13.97	2.43	(9.19, 18.73)	
$p(\beta_{00}\vert y)_{N/t_4}$	13.78	2.56	(8.74, 18.87)	
$p(\beta_{00}\vert y)_{t_4/t_4}$	13.93	2.42	(9.09, 18.65)	
$p(\beta_{00}\vert y)_{t_4/t_3}$	13.96	2.42	(9.11, 18.68)	
Mean Rate				
$p(\beta_{10}\vert y)_{N/N}$	-5.23	0.493	(-6.21, -4.26)	
$p(\beta_{10}\vert y)_{t_4/N}$	-5.21	0.493	(-6.16, -4.26)	
$p(\beta_{10}\vert y)_{N/t_4}$	-5.20	0.505	(-6.21, -4.20)	
$p(\beta_{10}\vert y)_{t_4/t_4}$	-5.20	0.489	(-6.15, -4.25)	
$p(\beta_{10}\vert y)_{t_4/t_3}$	-5.20	0.494	(-6.19, -4.24)	
MSPAC Coeff.				
$p(\beta_{11}\vert y)_{N/N}$	-0.082	0.042	(-0.165, -0.000)	.025
$p(\beta_{11}\vert y)_{t_4/N}$	-0.083	0.043	(-0.169, 0.001)	.026
$p(\beta_{11}\vert y)_{N/t_4}$	-0.088	0.042	(-0.171, -0.005)	.020
$p(\beta_{11}\vert y)_{t_4/t_4}$	-0.089	0.042	(-0.172, -0.004)	.021
$p(\beta_{11}\vert y)_{t_4/t_3}$	-0.090	0.043	(-0.174, -0.005)	.019

a. In the case of the MSPAC coefficient, the parameter of primary interest, we have included the posterior probability that β_{11} takes on values greater than 0 (e.g., $p(\beta_{11} > 0\vert y)_{N/N} = .025$).

0 [i.e., $p(\beta_{11} > 0 \,|\, \mathbf{y}) = 0.0246$]. To help interpret the value of the posterior mean for β_{11}, consider two children whose mothers differ by 30 percentage points in terms of MSPAC, which is similar to the range of MSPAC values in our sample. In this case, the expected difference in rates of change in enactive gesturing for two such children would be $|-0.082 \times 30|$, or approximately 2.5 percentage points per month.

We now reanalyze the data under t_4 assumptions at level 1, while retaining normality assumptions at level 2 (i.e., t_4/N). In growth modeling applications, employing heavy–tailed distributional assumptions at level 1 helps produce robust summaries of the data for each individual in a sample. Hence, the ensuing results for β_{11}, for example, will in effect be based on the relationship between an ensemble of robust growth parameter estimates for the children in our sample and the predictor MSPAC.

As can be seen in Table 2.1, the t_4/N analysis results in fairly minor changes in the results for the fixed effects. In particular, we see that the posterior mean for β_{11} takes on a slightly larger negative value. It can also be seen that there is a slight widening of the 95% interval for β_{11}. Although the upper boundary of the interval now includes 0, note that the marginal posterior probability that β_{11} exceeds 0 is extremely close to the value obtained in the N/N analysis (i.e., .026).

As noted earlier, in employing the scale mixture of normals representation of the t at level 1, each observation has a corresponding weight parameter (i.e., ω_{ti}). Similar to the use of robust regression techniques such as biweighting, where the final values of the weights for the observations in a data set can be used to identify outliers, the posterior means or medians of the ω_{ti} can be used to help identify extreme level–1 observations. In the t_4/N analysis, the resulting posterior means of the ω_{ti} range from 0.59 to 1.22. The smallest posterior mean is associated with the 16–month observation for child 1; the posterior means of the weight parameters connected with the 12, 20, and 24 month observations in this child's time series all exceed values of 1 (see Fig. 2.1). Under N/N distributional assumptions, the posterior mean of the growth rate for child 1 ($\pi_{1(1)}$) is -4.46. Downweighting the second observation in this child's time series results in a posterior mean for $\pi_{1(1)}$ that is slightly more negative (-4.55).

In this application, reanalyzing the data under heavy tails at level 1 results in small amounts of change in the summaries of the data for each child, and, in turn, little change in the results for the fixed effects. Thus, this analysis provides us with a certain amount of comfort. We have some assurance that the results concerning the relationship between MSPAC and rate of change are not being unduly influenced by one or two extreme outcome values. Examples in which level–1 outliers impact the results of fixed effects are treated later.

We now turn our attention to the issue of level 2 outliers and employ t_4 assumptions at level 2. As can be seen in Table 2.1, under N/t_4 assumptions and t_4/t_4 assumptions, the posterior mean of β_{11} takes on a value that is appreciably more negative. In addition, we see a downward shift in the 95% interval for β_{11}. To help understand this change, recall that in the scale mixture of normals representation of the t at level 2, each level–2 unit has a corresponding weight parameter (i.e., q_i; see Equations 2.8 and 2.9). In the N/t_4 analysis, note that the posterior mean of the level–2 weight parameter for child 9 takes on a value of 0.84, whereas the posterior means of the weight parameters for the other 8 children take on values ranging from 0.94 to 1.12. In the plot of LS rate of change estimates versus MSPAC values (Fig. 2.2), we see that child 9 deviates somewhat from the overall pattern. Specifically, although child 9's MSPAC value is fairly small, her rate of decline in enactive gesturing is fairly steep compared with other children with low MSPAC values. Thus, under normality assumptions at level 2, child 9, to some extent, pulls the level-2 fit toward her, which, in turn, has the effect of slightly dampening the magnitude of the posterior mean of β_{11}. Under t_4 level–2 assumptions, the pull exerted by child 9 is lessened to some extent.

Employing t_3 level–2 assumptions further reduces the influence of child 9 on the fit. The posterior mean of the level–2 weight parameter for child 9 drops to a value of 0.81 in the t_4/t_3 analysis, and the posterior mean for β_{11} takes on a value of -0.090.

The heavy–tailed level–2 analyses point to a somewhat stronger relationship between rate of decline in enactive gesturing and MSPAC. Practically speaking, however, under both normal and heavy–tailed level–2 assumptions, the results lead to fairly similar conclusions concerning the relationship between mother's speech about the child and rates of change in enactive gesturing. With only 9 children in our sample, the use of t assumptions at level 2 provides us with some assurance that our inferences concerning β_{11} are not being unduly influenced by a child whose rate of decline is unusually fast or slow. For examples in which level–2 outliers substantially affect conclusions regarding fixed effects of interest, see Seltzer (1993) for a growth modeling application, and Carlin (1992) for a meta–analysis application.

ILLUSTRATIVE EXAMPLE 2

Much of the literature on employing t distributional assumptions in HM settings has focused on the use of the t at level 2. A goal of this section is

TABLE 2.2

Example 2: Posterior distributions of the fixed effect for mother's speech
under $N/N, N/t_4, t_4/N$, and t_4/t_4 level-1/level-2 distributional
assumptions.

MSPAC Coeff.(β_{11})	Mean	SD	95% Interval	$p(\beta_{11} > 0 \mid y)$
$p(\beta_{11} \mid y)_{N/N}$	-0.057	0.041	(-0.139, 0.023)	.076
$p(\beta_{11} \mid y)_{N/t_4}$	-0.058	0.041	(-0.139, 0.022)	.073
$p(\beta_{11} \mid y)_{t_4/N}$	-0.085	0.041	(-0.163, -0.003)	.022
$p(\beta_{11} \mid y)_{t_4/t_4}$	-0.089	0.042	(-0.169, -0.005)	.020

to highlight the important role that the use of t distributional assumptions
at level 1 can play in sensitivity analysis. In particular, we wish to bring
to light situations where the impact of level–1 outliers on results for fixed
effects of interest can go undetected when t distributional assumptions are
employed at level 2, but normality assumptions are retained at level 1.

To help illustrate this point, we change child 3's EG value at age 16
months to a value of 90. Examining child 3's time series (see Fig. 2.1),
it can be seen that such an increase will result in a larger negative OLS
estimate of the rate of change for child 3 – that is, an estimate suggesting
a steeper rate of decline (cf. Figs. 2.2 and 2.3).

Under normality assumptions at levels 1 and 2, we see in Table 2.2
that the marginal posterior mean of β_{11} (i.e., the MSPAC fixed effect)
takes on a substantially smaller negative value (-0.057) compared with the
results from the previous set of analyses. In addition, we see that the upper
boundary of the resulting 95% interval comfortably includes a value of 0;
more specifically, $p(\beta_{11} > 0 \mid \mathbf{y}) = 0.075$.

This difference in results is readily grasped when we compare Figs. 2.2
and 2.3. Note that child 3's MSPAC value is the smallest in the sample.
Thus, when rate of change is regressed on MSPAC, child 3 will exert an
appreciable amount of leverage on the resulting fit. In Fig. 2.2, we see that
child 3 is located in the extreme upper left-hand corner of the plot. In Fig.
2.3, however, child 3 is now shifted downward. The overall pattern in Fig.
2.3 suggests that the relationship between rate of change and MSPAC is
flatter (i.e., less negative) compared with Fig. 2.2.

When we retain normality assumptions at level 1 and employ t_4
assumptions at level 2, we see in Table 2.2 that there is virtually no
change in the results for β_{11}. In the N/t_4 analysis, the posterior means
of the level–2 weight parameters range from 0.95 to 1.09. In particular, the
posterior mean of the weight parameter for child 3 is 1.06.

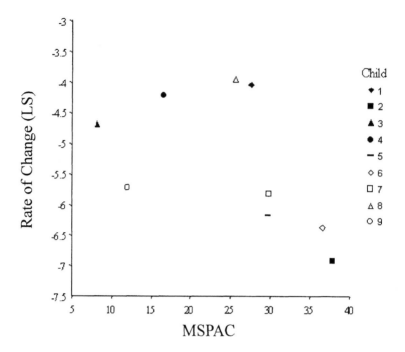

FIG. 2.3. Least squares estimates of rates of change in enactive gesturing versus level of mother's speech about the child for Illustrative Example 2.

We next employ t_4 assumptions at level 1 while retaining normality assumptions at level 2. The marginal posterior mean of β_{11} now takes on an appreciably larger negative value (-0.085), and the upper boundary of the 95% interval now lies below a value of 0. In addition, we see that $p(\beta_{11} > 0 \,|\, \mathbf{y}) = 0.022$.

This change in results stems from the fact that the 16–month observation for child 3 is downweighted substantially in this analysis. The posterior means of the level–1 weight parameters corresponding to the 12–, 16–, 20–, and 24–month observations for child 3 take on values of 1.00, 0.17, 0.95, and 1.11, respectively. This weighting scheme essentially returns us to a situation where the overall pattern of rate of change versus MSPAC resembles that in Fig. 2.2. Hence, the resulting value for the posterior mean of β_{11} in this analysis is extremely similar to the values that we obtained in the previous set of analyses (cf. Table 2.1).

It is also instructive to compare the marginal posterior means of the rate of change for child 3 ($\pi_{1(3)}$) obtained in the N/N and t_4/N analyses. In the N/N analysis, where each observation receives equal weight, we obtain a posterior mean of -4.53. However, in the t_4/N analysis, the posterior mean takes on a value that is substantially less negative (-3.52).

Reanalyzing the Data Under Varying Degrees of Heavy Tailedness

When answers that we obtain under heavy–tailed distributional assumptions differ appreciably from answers obtained under normality assumptions, it is valuable to study how answers change when we vary the degree of heavy tailedness. We now continue with our current example and illustrate this strategy employing a grid of values for ν_1.

In the literature, one finds that degrees of freedom parameters are sometimes fixed at values as small as 1. Because t distributions with 2 degrees of freedom or less have infinite variance, and because assumptions of infinite variance at levels 1 or 2 do not seem to be sensible in the context of our example, we focus on values of ν_1 of 3 or more. In these analyses, we fix ν_2 at a value of 4.

As can be seen in Table 2.3, the posterior means for β_{11} under t_3 level–1 assumptions and under normal level–1 assumptions differ by approximately 0.033 points. As ν_1 increases, we see that the posterior mean for β_{11} gradually becomes less negative, and that the upper boundary of the 95% credible interval takes on increasingly larger values. Note, however, that even when ν_1 is fixed at values of 11, 15, and 20, the results that we obtain still differ appreciably from those obtained under level–1 normality assumptions.

In this example, we focused on the presence of extreme level–1 observations nested within level–2 units that are leverage points (i.e., level–2 units that take on relatively large or small values in the space of the level–2 predictors). Note, however, that the impact of level–1 outliers on the estimation of fixed effects is decidedly not confined to such situations (see, e.g., Seltzer et al., in press). A key point is that estimates of fixed effects of interest based on a set of robust summaries of the level–1 data can differ substantially from estimates based on a set of nonrobust summaries.

Treating Degrees of Freedom Parameters as Unknowns

In situations where answers change appreciably across a range of values specified for a degrees of freedom parameter, it becomes important to try to calculate the marginal posterior distribution of that parameter [e.g., $p(\nu_1 \mid \mathbf{y})$]. In the context of our current example, this would provide a means of assessing the relative plausibility of various values for ν_1 given the data

TABLE 2.3

Posterior means, standard deviations and intervals for the MSPAC coefficient in Example 2 (a) fixing the degrees of freedom at level-1 (ν_1) along a grid of values, and (b) treating ν_1 as unknown. In the analyses in which ν_1 is treated as unknown, three priors were employed; these three analyses reflect increasing amounts of prior probability placed on approximate Gaussian tails at level-1 (see the text below for details).

	Mean	SD	95% Interval	Prob.> 0
Results for the MSPAC Coeff. (β_{11})				
Treating ν_1 fixed along a grid of values:				
t_3/t_4	-0.091	0.041	(-0.171, -0.007)	.018
t_4/t_4	-0.089	0.042	(-0.169, -0.005)	.020
t_7/t_4	-0.083	0.041	(-0.162, -0.001)	.024
t_{11}/t_4	-0.077	0.042	(-0.158, 0.008)	.034
t_{15}/t_4	-0.073	0.042	(-0.155, 0.009)	.039
t_{20}/t_4	-0.070	0.041	(-0.152, 0.012)	.045
t_{30}/t_4	-0.066	0.042	(-0.147, 0.017)	.056
N/t_4	-0.058	0.041	(-0.139, 0.022)	.073
Treating ν_1 as unknown [a] :				
Prior 1	-0.085	0.042	(-0.168, -0.002)	.023
Prior 2	-0.083	0.042	(-0.165, 0.000)	.026
Prior 3	-0.081	0.042	(-0.164, 0.002)	.028

Note. The degrees of freedom at level-2 (ν_2) is fixed at a value of 4.

at hand and based on prior information or beliefs concerning ν_1. As will be seen, the process of using the data and any available prior information to estimate ν_1 gives rise to estimates for parameters of interest in HMs (e.g., β_{11}) that are analogous to adaptive robust estimates.

In using BUGS to conduct analyses of this kind, we must treat degrees of freedom parameters as discrete variables that take on various prespecified values (for alternative strategies, see Seltzer et al., in press, and Draper, in press). The values may be equally spaced (e.g., $\nu_1 = 3, 4, 5, \ldots, 98, 99, 100$), or, the spacing can vary (e.g., 2, 4, 6, 8, 10, 12, 15, 20, 30, 50; see Spiegelhalter, Thomas, et al., 1996b, p. 35). In using this approach, however, we have found that the autocorrelation among the values generated for degrees of freedom parameters in runs of the Gibbs sampler can be quite high. In such situations, the Gibbs sampler can "get stuck" for an appreciable number of iterations at the same value for ν_1 or ν_2 (see also Spiegelhalter, Best, et al., 1996a, p. 37).

This has led us to consider possible reparameterizations. In many applications of MCMC, one often encounters parameterizations that involve log transformations or reciprocal transformations of various parameters in a given model. In the case of degrees of freedom parameters, Gelman et al. (1995), Liu (1995), and Gelman and Meng (1996) employed parameterizations that involved the reciprocal of degrees of freedom parameters (e.g., $1/\nu$). In the case of our HM application, we found that the use of this parameterization in the WinBUGS environment resulted in substantial reductions in autocorrelation, thus helping to alleviate the kinds of problems previously discussed. The beauty of techniques such as the Gibbs sampler is that we can simulate the marginal posterior for ν_1, for example, simply by inverting the values generated for $1/\nu_1$.

Thus, in our analyses, we let $1/\nu_1$ take on the following values: .01, .02, .03, $\ldots$, .31, .32, .33. Note importantly that a value of $1/\nu_1 = .01$ corresponds to a value of $\nu_1 = 100$, and that a value of $1/\nu_1 = .33$ corresponds to a value of $\nu_1 \approx 3$.

With regard to specifying a prior for $1/\nu_1$, we are not aware of any published or unpublished work on enactive gesturing from which we might be able to obtain prior information concerning level–1 tail behavior. However, more generally, we do know that there is a tendency for data in the social and behavioral sciences to be fairly noisy, especially in cases where the data are based on observations in field settings (e.g., home environments, classrooms). Even in the physical sciences, it is quite common to encounter sets of measurements that contain outliers or that exhibit fairly heavy tails (see Draper, in press, p. 97; see also Gelman et al., 1995, pp. 166–167).

In our analyses, we consider three priors for $1/\nu_1$. The first prior places equal probability on each of the prespecified values for $1/\nu_1$. Translating back to ν_1, this in effect corresponds to placing a fairly high degree of prior probability on heavy tails at level 1 [i.e., $p(3 < \nu_1 < 5) \approx .39$], and a fairly small amount of prior probability on tails that are roughly Gaussian [i.e., $p(20 \leq \nu_1 \leq 100) \approx .15$]. To see this, note that values for $1/\nu_1$ that range from .21 to .33 (i.e., .21, .22, ..., .32, .33) correspond to values of ν_1 that lie between 3 and 5, whereas values for $1/\nu_1$ that range from .01 to .05 correspond to values of 100, 50, 33.33, 25, and 20 for ν_1. Thus, when we place equal prior probability on each of the prespecified values for $1/\nu_1$, we are, in fact, placing a large amount of prior probability on heavy-tailed level-1 assumptions. The second prior (Prior 2) that we employ in effect places approximately equal amounts of prior probability on heavy level-1 tails and roughly Gaussian tails [$p(3 < \nu_1 < 5) = .24$; $p(20 \leq \nu_1 \leq 100) \approx .25$]. The third prior (Prior 3) places a prior probability of .375 on approximately Gaussian tails and a prior probability of .225 on heavy tails.

In an analysis based on our first prior (Prior 1), the resulting posterior mean for β_{11} is -0.085, which lies between the values that we obtained with ν_1 fixed at values of 4 and 7, respectively (see Table 2.3 and Fig. 2.4). We also see that the upper boundary of the 95% interval lies below a value of 0. Using Priors 2 and 3, we see that as we increase the amount of prior probability placed on roughly Gaussian tails, the posterior mean for β_{11} takes on slightly less negative values, and the upper boundary of the 95% interval includes values slightly larger than 0. Note that the marginal posterior probability that β_{11} exceeds 0 [i.e., $p(\beta_{11} > 0 \,|\, \mathbf{y})$] differs by an extremely small amount across the three analyses.

This pattern of results can be readily grasped by considering the following. The top half of Table 2.3 reports the results that we obtain for β_{11} when we condition on a series of different values for ν_1 [e.g., $p(\beta_{11} \,|\, \mathbf{y}, \nu_1 = 3)$, $p(\beta_{11} \,|\, \mathbf{y}, \nu_1 = 4)$, etc.]. When ν_1 is treated as an unknown, the resulting marginal posterior distribution of β_{11} [$p(\beta_{11} \,|\, \mathbf{y})$] can be viewed as a weighted average of a set of conditional posteriors for β_{11} (i.e., conditional on different possible values for ν_1) where weights are supplied by $p(\nu_1 \,|\, \mathbf{y})$. Note that under Priors 1, 2 and 3, $p(3 < \nu_1 < 7 \,|\, \mathbf{y})$ takes on values of .71, .61, and .58, respectively, whereas $p(20 \leq \nu_1 \leq 100 \,|\, \mathbf{y})$, for example, takes on values of .05, .08, and .16. Thus, in all three analyses, extremely large amounts of weight are placed on conditional posteriors for β_{11} that are conditional on fairly small values of ν_1.

Even with only 35 level-1 observations, the data in this example are fairly informative regarding tail behavior at level 1. As the number of level-1 observations in a data set increase, inferences concerning ν_1 will tend to be increasingly insensitive to choice of priors for ν_1.

One can also treat ν_2 as an unknown in HM analyses. When the

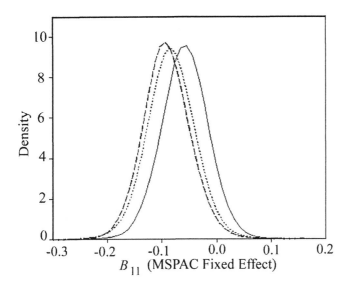

FIG. 2.4.   Illustrative Example 2: Marginal posterior distribution of the MSPAC fixed effect under (a) N/t_4 assumptions (solid line); (b) t_3/t_4 assumptions (dashed line); and (c) t/t_4 assumptions (dotted line) where the degrees of freedom at level 1 is treated as unknown. For (c) we employ a prior that places approximately equal amounts of prior probability on heavy level-1 tails and roughly Gaussian tails.

number of level–2 units in an application is small, however, the data will tend to provide very little information regarding level–2 tail behavior, and inferences may be highly sensitive to choice of prior for ν_2.

DISCUSSION

A key aim of this chapter is to illustrate the value of conducting sensitivity analyses via the use of t distributional assumptions at levels 1 and 2 of HMs. Our strategy is based on the scale mixture of normals representation of the t. A strategy of this kind can help us identify extreme level-1 and level–2 units, and study the sensitivity of inferences concerning fixed and random effects of interest to such units.

Examining the posterior means of the level-1 and level-2 weight parameters in this formulation provides a useful supplement to available diagnostics for identifying outliers in HM settings (e.g., inspecting plots of residuals). In general, in conducting analyses under normality assumptions, fitted surfaces can be pulled substantially toward extreme cases; as Hogg (1979) notes, outliers may be masked in plots of residuals in such situations.

For HM settings in which vectors of level–1 parameters are assumed to vary across level–2 units (e.g., initial status and individual growth rate parameters in growth modeling appplications; site means and treatment effects in multisite evaluation studies), Seltzer et al. (1996) illustrated the use of multivariate t (MVT) distributional assumptions at level 2 (see also, Wakefield, Smith, Racine-Poon, & Gelfand, 1994). Although one can specify MVN distributional assumptions in WinBUGS, MVT assumptions cannot as of yet be specified. However, in those situations in which the same set of predictors is specified in each level–2 equation, we have found that the use of univariate t assumptions in each equation often yields results that are similar to those obtained under MVT assumptions.

When different predictor sets are used in the level–2 equations in an HM, Bryk and Raudenbush (1992) warned that mispecification in one equation can distort estimates in another equation. The use of univariate t distributions at level–2, which in essence sets level–2 covariance terms to 0, provides us with a specification check akin to the specification check outlined by Bryk and Raudenbush (1992, p. 216).

One can easily specify HMs consisting of three or more levels in WinBUGS, and t distributional assumptions can be employed at any level using the scale mixture of normals approach outlined in this chapter. WinBUGS also enables one to employ an array of link functions at level 1 (e.g., logit, probit, log, complementary log–log).

In this chapter, we focus on level–1 units that take on unusually large or small values with respect to Y, and on level–2 units that are extreme with respect to level–1 parameters that are treated as random (e.g., individuals with unusually rapid or slow rates of change; sites at which an intervention has been unusually successful). Drawing sound inferences, particularly in small–sample settings, of course requires more than attending to level–1 and level–2 outliers. For example, as is well known, cases that are extreme in the space of the predictors in a model can exert a high degree of influence on one's results. Thus, a strategy that we recommend, particularly in small–sample settings, entails sequentially setting aside one level–2 unit at a time from one's sample, refitting one's model, and examining changes in results for parameters of interest.

Another important approach to model checking is posterior predictive model checking, which is discussed in detail in Rubin (1981, 1984), Gelman et al. (1995), Belin and Rubin (1995), and Gelman and Meng (1996). Posterior predictive model checking, which can be implemented quite naturally using MCMC, provides a means of identifying important discrepancies between the data one is analyzing and one's model. Finally Gelfand, Dey, and Chang (1992) presented a cross–validation approach that combined the logic of "leave–one–out analyses" with that of posterior predictive model checking (see also Draper, 1998).

ACKNOWLEDGMENTS

We wish to thank the Center of Research on Evaluation, Standards, and Student Testing at UCLA and the National Science Foundation (Grant SBR-9422901) for their generous support. We also wish to thank Patricia Smiley for permission to use the data from her study on enactive gesturing in children to illustrate the methods presented in this chapter.

ENDNOTES

1. Smiley et al. also included children's mean length utterance, which is a time–varying covariate, in their level–1 model. This covariate was not statistically significant, and results concerning the relationship between MSPAC and children's rates of change in enactive gesturing were extremely similar whether one included this covariate in the analysis or not.

2. In the case of HMs in which none of the level–2 variances is constrained to be 0, least squares (LS) estimates of individual growth parameters can be obtained simply by regressing the set of outcome values for a person on the set of level–1 predictor values for that person. However, when, for example, the variance connected with a particular facet of growth is constrained to be 0, or when the level–1 model contains a time–varying covariate that is treated as fixed, computing LS estimates of individual growth parameters becomes more complex. In such cases, we find it extremely convenient to obtain LS estimates using residual files produced by the HLM program. In the case of our model, for example, HLM computes a fitted value of the growth rate for each child (i.e., $\hat{\beta}_{10} + \hat{\beta}_{11} MSPAC_i$), and an EB (shrinkage) estimate (U_{1i}^*) and an LS estimate ($\hat{U}_{1i}$) of the random effect for each child. Adding the EB estimates of the random effects to the corresponding fitted values yields EB estimates of the π_{1i} for the children in our sample, and adding the LS estimates of the random effects to the fitted values yields LS estimates of the π_{1i}.

3. The Gibbs sampler, in effect, walks through or traverses the joint

posterior distribution of all unknowns in one's model (see, e.g., Gelman, Bois, & Jiang, 1996). The relative frequency with which the Gibbs sampler visits various regions of the joint posterior is proportional to the joint posterior density of each region. One thing that is important to attend to in working with the Gibbs sampler is that in some cases, the values sampled in successive iterations for particular parameters may be highly correlated. In extreme cases, the Gibbs sampler may become "stuck" for a number of iterations in certain regions of the parameter space.

To help diagnose possible problems, it is important to run multiple sequences (chains) of the Gibbs sampler (see Gelman & Rubin, 1992). Thus, for each analysis, we ran two chains of the Gibbs sampler using different starting values for $1/\tau_{11}$ and $1/\sigma^2$ and using different seeds. (Note that in WinBUGS, we work with reciprocals of variance components, i.e., precisions.) For one set of starting values, we used the reciprocals of the REML estimates for τ_{11} and σ^2 that we obtained using the HLM program. The second set of starting values was based on results from the first chain; specifically, we inverted the .025 quantiles of the marginal posteriors for τ_{11} and σ^2. After a "burn–in" or "warm–up" period of 2,000 iterations, we ran each chain for an additional $M = 40,000$ iterations. Setting aside the results from the burn–in period, we then compared results (e.g., the empirical distributions for all parameters in the model) across the two chains. For each analysis, the chains that we ran yielded highly similar results. In addition, we assessed convergence through the use of trace plots and autocorrelation function plots, which can be obatined in both WinBUGS and CODA, and through the use of a diagnostic procedure developed by Raftery and Lewis (1996), which is available in CODA. For each analysis, the marginal posteriors that we report are based on deviates generated in the first chain of $M = 40,000$ iterations.

4. In each of our analyses we placed diffuse priors on the fixed effects. In particular, for each fixed effect, we specified a normal prior with extremely small precision. In settings in which our samples consist of small numbers of level–2 units, care must be used in specifying priors for level–2 variance components. For example, certain choices can result in intervals for fixed effects that are too liberal with respect to coverage, and other choices can result in intervals that are too conservative. In this chapter, we use a strategy discussed and illustrated in papers by Seltzer et al. (1996, in press); the results of a simulation study reported in Seltzer et al. (in press) indicate that this approach tends to produce intervals for fixed effects with good coverage properties. This strategy involves the use of data–based priors for the variance components. The basic idea is to specify priors for the variance components that are fairly diffuse, but with modes approximately equal to

REML estimates of the variance components.

In all analyses, we placed gamma priors on $1/\sigma^2$ and $1/\tau_{11}$. The gamma distribution has two parameters: $Gamma(a, b)$, where a is a shape parameter and b is an inverse scale parameter. Following the approach discussed in Seltzer et al. (1996, in press), we set a equal to a value of 1.5 in all of our analyses. In terms of priors for $1/\tau_{11}$, under the assumption of normality at level 2, b was set to the following value: $b = \hat{\tau}_{11} \times (a + 1)$, where $\hat{\tau}_{11}$ is the REML estimate of τ_{11}. Under t level–2 assumptions with ν_2 fixed at a particular value, we employed three different values for b as a kind of sensitivity analysis: (a) the value of b used under normality assumptions at level 2, which we refer to as b_N; (b) $b_{tI} = [(\nu_2 - 2)/\nu_2] \times b_N$; and (c) $b_{tII} = (b_N + b_{tI})/2$. Little material difference was found in our results for parameters of interest under these three different choices for b. The results that we present in the chapter are based on b_{tII}. In terms of priors for $1/\sigma^2$, we used a similar procedure in specifying values for b, with the REML estimate of σ^2 (i.e., $\hat{\sigma}^2$) replacing $\hat{\tau}_{11}$, and ν_1 replacing ν_2.

REFERENCES

Belin, T., & Rubin, D. (1995). The analysis of repeated–measures data on schizophrenic reaction times using mixture models. *Statistics in Medicine, 14*, 747–768.

Best, N., Cowles, M., & Vines, K. (1996). *CODA: Convergence diagnosis and output analysis software for Gibbs sampling output, [Version 0.30]*. MRC Biostatistics Unit, Cambridge University, UK.

Box, G. E. P. (1979). Robustness in the strategy of scientific model building. In R. Launer & G. Wilkinson (Eds.), *Robustness in statistics* (pp. 201–236). New York: Academic Press.

Box, G. E. P. (1980). Sampling and Bayes inference in scientific modeling and robustness (with discussion). *Journal of the Royal Statistical Society, 143* (Series A), 383–430.

Box, G. E. P., & Tiao, G. (1973). *Bayesian inference in statistical analysis*. Reading, MA: Addison-Wesley.

Bryk, A. S., & Raudenbush, S. W. (1992). *Hierarchical linear models: Applications and data analysis methods*. Newbury Park, CA: Sage.

Bryk, A. S., Raudenbush, S. W., & Congdon (1996). *HLM: Hierarchical linear and non–linear modeling with the HLM/2L and HLM/3L programs*. Chicago: Scientific Software International.

Carlin, B. (1992). Comment on Morris and Normand (1992). In J. M.

Bernardo, J. O. Berger, A. P. Dawid, & A. S. Smith (Eds.), *Bayesian statistics 4* (pp. 336–338). New York: Oxford University Press.

Carlin, B., & Louis, T. (1996). *Bayes and empirical Bayes methods for data analysis.* London: Chapman & Hall.

Dempster, A. P., Laird, N. M., & Rubin, D. B. (1980). Iteratively reweighted least squares. In P. R. Krishnaiah (Ed.), *Multivariate analysis V* (pp. 35–57). Amsterdam: North-Holland.

Draper, D. (1995). Inference and hierarchical modeling in the social sciences. *Journal of Educational and Behavioral Statistics, 20,* 115–147.

Draper, D. (1998). Comment on Hodges (1998). *Journal of the Royal Statistical Society, 60* (Series B), 527–528.

Draper, D. (in press). *Bayesian hierarchical modeling.* Springer–Verlag.

Gelfand, A., Dey, D., & Chang, H. (1992). Model determination using predictive distributions with implementation via sampling–based methods. In J. Bernardo, J. Berger, A. Dawid, & A. Smith (Eds.), *Bayesian statistics 4* (pp. 147–167). New York: Oxford University Press.

Gelfand, A., Hills, S., Racine-Poon, A., & Smith, A. (1990). Illustration of Bayesian inference in normal data models using Gibbs sampling. *Journal of the American Statistical Association, 85,* 972–985.

Gelfand, A. E. & Smith, A. F. M. (1990). Sampling based approaches to calculating marginal densities. *Journal of the American Statistical Association, 85,* 398–409.

Gelman, A., Bois, F., & Jiang, J. (1996). Physiological pharmokinetic analysis using population modeling and informative prior distributions. *Journal of the American Statistical Association, 91,* 1400–1412.

Gelman, A., Carlin, J., Stern, H., & Rubin, D. (1995). *Bayesian data analysis.* London: Chapman & Hall.

Gelman, A., & Meng, X. (1996). Model checking and model improvement. In W. Gilks, S. Richardson, & D. Spiegelhalter (Eds.), *Markov Chain Monte Carlo in practice* (pp. 189–201). London: Chapman & Hall.

Gelman, A., & Rubin, D. B. (1992). Inference from iterative simulation using multiple sequences. *Statistical Science, 7,* 457–472.

Gilks, W., Richardson, S., & Spiegelhalter, D. (1996). *Markov Chain Monte Carlo in practice.* London: Chapman & Hall.

Hogg, R. (1979). Statistical robustness: One view of its use in applications today. *American Statistician, 33,* 108–115.

Lange, K., Little, R., & Taylor, J. (1989). Robust statistical modeling using the *t* distribution. *Journal of the American Statistical Association, 84,* 881–896.

Liu, C. (1995). Missing data imputation using the t distribution. *Journal of Multivariate Analysis, 53,* 139–158.

Mosteller, F., & Tukey, J. (1977). *Data analysis and regression.* Reading, MA: Addison-Wesley.

Rachman-Moore, D., & Wolfe, R. (1984). Robust analysis of a nonlinear model for multilevel educational survey data. *Journal of Educational Statistics, 9,* 277–293.

Raftery, A., & Lewis, S. (1996). Implementing MCMC. In W. Gilks, S. Richardson, & D. Spiegelhalter (Eds.), *Markov Chain Monte Carlo in practice* (pp. 115–130). London: Chapman & Hall.

Rubin, D. B. (1981). Estimation in parallel randomized experiments. *Journal of Educational Statistics, 6,* 377–400.

Rubin, D. B. (1984). Bayesianly justifiable and relevant frequency calculations for the applied statistician. *Annals of Statistics, 12,* 1151–1172.

Seltzer, M. (1993). Sensitivity analysis for fixed effects in the hierarchical model: A Gibbs sampling approach. *Journal of Educational Statistics, 18,* 207–235.

Seltzer, M., Novak, J., Choi, K., & Lim, N. (in press). Sensitivity analysis for hierarchical models employing t level-1 assumptions. *Journal of Educational and Behavioral Statistics.*

Seltzer, M., Wong, W. H., & Bryk, A. (1996). Bayesian inference in applications of hierarchical models: Issues and methods. *Journal of Educational and Behavioral Statistics, 21,* 131–167.

Smiley, P., Greene, J., Seltzer, M., & Choi, K. (2000). *Knowledge of self and other in interaction: Development of requesting in 12- to 30-month-olds and its relation to language input.* Manuscript in preparation.

Spiegelhalter, D., Best, N., Gilks, W., & Inskip, H. (1996a). Hepatitis B: A case study in MCMC methods. In W. Gilks, S. Richardson, & D. Spiegelhalter (Eds.), *Markov Chain Monte Carlo in practice* (pp. 21–43). London: Chapman & Hall.

Spiegelhalter, D., Thomas, A., & Best, N. (2000). *WinBUGS [Version 1.3].* MRC Biostatistics Unit, Cambridge University, UK.

Spiegelhalter, D., Thomas, A., Best, N., & Gilks, W. (1996b). *BUGS examples (Vol. 1, Version 0.50).* MRC Biostatistics Unit, Cambridge University.

Spiegelhalter, D., Thomas, A., Best, N. & Gilks, W. (1996c). *BUGS examples volume 2, Version 0.50.* MRC Biostatistics Unit, Cambridge University.

Tanner, M. (1996). *Tools for statistical inference* (3rd ed.). New York: Springer-Verlag.

Tanner, M. A., & Wong, W. H. (1987). The calculation of posterior distributions by data augmentation. *Journal of the American Statistical Association, 82,* 528–550.

Wakefield, J., Smith, A. F. M., Racine-Poon, A., & Gelfand, A. (1994). Bayesian analysis of linear and nonlinear population models using the Gibbs sampler. *Applied Statistics, 43,* 201–221.

West, M. (1984). Outlier models and prior distributions in Bayesian linear regression. *Journal of the Royal Statistical Society* (Series B), 46, 431–439.

3

Two-level Mean and Covariance Structures: Maximum Likelihood via an EM Algorithm

Peter M. Bentler
*University of California,
Los Angeles*

Jiajuan Liang
*University of New Haven,
School of Business*

Methodologies for the analysis of two-level structural equation models (SEM for simplicity) have been proposed by a number of authors. These methodologies are usually suitable for some specific formulation of two-level SEM. Goldstein and McDonald (1988) proposed a general model for analysis of multilevel data that includes two-level SEM as its special case. McDonald and Goldstein (1989) proposed a general treatment for maximum likelihood (ML) analysis of two-level SEM. The importance of ML lies in its asymptotic optimality, that is, an estimator with the smallest standard error, meeting the Cramér-Rao lower bound. For an unbalanced design of a sample, McDonald and Goldstein's (1989) algorithm seems to be computationally burdensome because a large number of inverse matrices have to be computed to obtain the maximum likelihood estimates (MLE) of model parameters. Muthén (1994) summarized the techniques in several papers in which he developed the so-called MUML analysis for two-level SEM. He implemented a pseudobalanced solution for unbalanced sample designs. MUML analysis of two-level SEM is a kind of approximate ML analysis (Hox, 2000). Raudenbush (1995) used a balanced complete data routine to show how ML analysis on two-level SEM with unbalanced designs can be done by available software. Lee (1990) proposed a simple formulation of two-level SEM and obtained general asymptotic properties of MLE for model parameters. Lee and Poon (1998) proposed a treatment for exact ML analysis of two-level SEM via EM type algorithms. An advantage of Lee and Poon's (1998) method is that it is applicable to arbitrary (balanced and unbalanced) sample designs and their algorithm turns out to converge fast.

The purpose of this chapter is to generalize Lee and Poon's (1998) model and their treatment for two-level SEM without a mean structure to the case with both mean and covariance structures and to provide a new computational approach to the estimation of parameters. The two-level SEM is defined by (Lee, 1990; Lee & Poon, 1998) as

$$x_{gi} = v_g + v_{gi}, \tag{3.1}$$

where x_{gi} is a vector of responses (observable random vector) from individual i (level-1 unit) nested in group g (level-2 unit). For example, x_{gi} may denote the measures of a student's achievements for several courses. A random sample of different students from different schools constitutes a random sample of two-level data, $\{x_{gi} : i = 1, \ldots, N_g; g = 1, \ldots, G\}$, say, that is, randomly choosing $N = \sum_{g=1}^{G} N_g$ level-1 units (students) from G randomly selected level-2 units (schools). In order to develop the ML analysis for Equation 3.1 with both mean and covariance structures via an EM algorithm, the following assumptions are necessary:

1. $\{v_g : g = 1, \ldots, G\}$ are i.i.d. (independently identically distributed) latent random vectors ($p \times 1$) varying only at level 2, and v_g has a normal distribution $N_p(\mu, \Sigma_B)$ with $\Sigma_B > 0$ (positive definite);

2. for each fixed g, $\{v_{gi} : i = 1, \ldots, N_g\}$ are i.i.d. latent random vectors ($p \times 1$) varying only at level 1, and $v_{gi} \sim N_p(0, \Sigma_{gW})$ with $\Sigma_{gW} > 0$ (Σ_{gW} may be different for $g = 1, \ldots, G$);

3. for each fixed g, v_g and v_{gi} ($i = 1, \ldots, N_g$) are uncorrelated (i.e., independent in the normal case);

4. for different level-2 indices $g_1 \neq g_2$, $\{v_{g_1 i} : i = 1, \ldots, N_{g_1}\}$ and $\{v_{g_2 i} : i = 1, \ldots, N_{g_2}\}$ are assumed to be independent.   Lee and Poon's (1998) treatment of ML analysis for Equation 3.1 is suitable for the case without a mean structure, that is, in assumption 1 where $\mu = 0$, or $v_g \sim N_p(0, \Sigma_B)$. When $\mu \neq 0$ and the components of μ can be considered as free parameters, μ could be estimated by the overall sample mean: $\bar{x} = \frac{1}{N} \sum_{g=1}^{G} \sum_{i=1}^{N_g} x_{gi}$. In this case, Lee and Poon's (1998) method for ML analysis of Equation 3.1 could still be approximately carried out

based on the centered two-level data

$$x_{gi} - \bar{x}: \quad i = 1, \ldots, N_g; \quad g = 1, \ldots, G. \tag{3.2}$$

The mean and covariance structures for Equation 3.1 are defined as

$$\mu = \mu(\theta), \quad \Sigma_B = \Sigma_B(\theta), \quad \Sigma_{gW} = \Sigma_{gW}(\theta), \tag{3.3}$$

where $\theta = (\theta_1, \ldots, \theta_r)'$ ($r \times 1$) is a parameter vector consisting of all interesting free parameters in Equation 3.1. For the mean and covariance structure (3.3), an ML analysis for Equation 3.1 based on the centered data (Equation 3.2) completely ignores the mean structure in Equation 3.3. This may bring about inaccuracy for the MLE of the parameter vector θ or it may result in poor model fit relative to degrees of freedom. Furthermore, smaller standard errors could be obtained for covariance structure parameters when a mean structure is imposed (Yung & Bentler, 1999). Therefore, particular methods for ML analysis of Equation 3.1 with mean and covariance structure (3.3) are necessary.

Equation 3.1 is a simple formulation of two-level SEM. It covers several interesting two-level mean and covariance structure models. For example, a two-level factor analysis model can be defined by (Muthén, 1994):

$$\begin{aligned} y_{gi} &= \nu + \Lambda \eta_{gi} + \epsilon_{gi}, \\ \eta_{gi} &= \alpha + \eta_{Bg} + \eta_{Wgi}, \end{aligned} \tag{3.4}$$

where y_{gi} is a vector of response, ν is a vector of intercepts, ϵ_{gi} is a vector of residuals, α is the overall expectation (grand mean), η_{Bg} is a random factor component capturing level-2 (e.g., organizational) effects, and η_{Wgi} is a random factor component varying over individual levels (level 1) within their organizations. Let $v_g = \nu + \Lambda \alpha + \Lambda \eta_{Bg}$ and $v_{gi} = \Lambda \eta_{Wgi} + \epsilon_{gi}$ in Equation 3.4. Then Equation 3.4 has the form of Equation 3.1 with $\mu = E(v_g) = \nu + \Lambda \alpha$ and $E(v_{gi}) = 0$ [assuming $E(\eta_{Bg}) = 0$, $E(\eta_{Wgi}) = 0$ and $E(\epsilon_{gi}) = 0$]. It is noted that $\mu = E(v_g) = \nu + \Lambda \alpha$ involves parameters from the matrix of factor loadings Λ. Therefore, Equation 3.4 can be formulated as Equation 3.1 with mean and covariance structure (3.3). In addition to two-level factor analysis models, a variety of other two-level SEM like those studied by Heck and Thomas (2000, chaps. 6–7) can also be formulated as Equation 3.1 with mean and covariance structure (3.3).

The main idea for deriving the ML analysis of Equation 3.1 via an EM algorithm is to formulate it as a missing data problem . Section 2 presents the details for deriving the EM algorithm. Section 3 gives some simplified formulas for computing the asymptotic standard errors of the MLE of model parameters and the chi-squared statistic for testing model fit. An ML analysis on a practical data set is demonstrated in Section 4 by

the proposed algorithm. Some concluding remarks on the convergence of
the proposed EM gradient algorithm and its possible further improvement
are given in the last section.

SECTION 2: THE EM GRADIENT ALGORITHM

Let $\{x_{gi} : i = 1, \ldots, N_g; \ g = 1, \ldots, G\}$ be a set of response vectors ($p \times 1$
random vectors) from Equation 3.1 and

$$x_g = \begin{pmatrix} x_{g1} \\ \vdots \\ x_{gN_g} \end{pmatrix}, \quad x = \begin{pmatrix} x_1 \\ \vdots \\ x_G \end{pmatrix}, \quad y_g = \begin{pmatrix} v_g \\ x_g \end{pmatrix}, \quad y = \begin{pmatrix} y_1 \\ \vdots \\ y_G \end{pmatrix},$$

(3.5)

and

$$\bar{x}_g = \frac{1}{N_g} \sum_{i=1}^{N_g} x_{gi}, \quad X_g = \begin{pmatrix} x_{g1}' \\ \vdots \\ x_{gN_g'} \end{pmatrix}, \quad X = \begin{pmatrix} X_1 \\ \vdots \\ X_G \end{pmatrix}. \quad (3.6)$$

To formulate Equation 3.1 as a missing data problem, we consider the y_g
in Equation 3.5 as the complete data from level-2 unit (group) g with a
missing value (random vector) v_g. For the purpose of deriving the EM
algorithm for computing the MLE of the model parameter vector θ ($r \times 1$),
we use an arbitrarily specified value of θ, say, θ^*, to express the $-2 \log ML$
function, say, $l(y, \theta^*)$, of the complete data $\{y_g : g = 1 \ldots, G\}$. Under the
assumptions 1 through 4 in Section 1, $l(y, \theta^*)$ can be written (except for
an additive constant) as:

$$\begin{aligned}
l(y, \theta^*) &= -2 \log ML \\
&= \sum_{g=1}^{G} N_g \left\{ \log |\Sigma_{gW}| + \mathrm{tr}\{\Sigma_{gW}^{*-1}[\frac{1}{N_g} \sum_{i=1}^{N_g} (x_{gi} - v_g)(x_{gi} - v_g)']\} \right\} \\
&\quad + G \left\{ \log |\Sigma_B^*| + \mathrm{tr}\{\Sigma_B^{*-1}[\frac{1}{G} \sum_{g=1}^{G} (v_g - \mu^*)(v_g - \mu^*)']\} \right\},
\end{aligned}$$

(3.7)

where $\mu^* = \mu(\theta^*)$, $\Sigma_{gW}^* = \Sigma_{gW}(\theta^*)$, $\Sigma_B^* = \Sigma_B(\theta^*)$ and the notation "tr"
denotes the "trace" of a matrix. The EM algorithm requires computation
of the E-step function

$$M(\theta^*|\theta) = E\{l(y, \theta^*)|x, \theta\}, \quad (3.8)$$

where both θ^* and θ are any two specified values of the same parameter
vector θ. See Dempster, Laird, and Rubin (1977) for the principle of the
EM algorithm.

Simplification of the E-step

The E-step function defined by Equation 3.8 requires computation of the conditional expectation

$$
\begin{aligned}
a_g &\overset{\text{def}}{=} E(v_g|x,\theta) = E(v_g|x_g,\theta) = \mu + a_{g0}, \\
C_g &\overset{\text{def}}{=} E(v_g v_g'|x,\theta) = E(v_g v_g'|x_g,\theta) \\
&= \Sigma_B - N_g\Sigma_B\Sigma_g^{-1}\Sigma_B + a_g a_g',
\end{aligned}
\tag{3.9}
$$

where the sign "$\overset{\text{def}}{=}$" means "defined as" and

$$
a_{g0} = N_g\Sigma_B\Sigma_g^{-1}(\bar{x}_g - \mu), \quad \Sigma_g = \Sigma_{gW} + N_g\Sigma_B.
\tag{3.10}
$$

Then we can write Equation 3.8 as

$$
\begin{aligned}
M(\theta^*|\theta) = &\sum_{g=1}^{G} N_g\left[\log|\Sigma_{gW}^*| + \text{tr}(\Sigma_{gW}^{*-1}S_{gW})\right] \\
&+ G\left[\log|\tilde{\Sigma}_B^*| + \text{tr}(\tilde{\Sigma}_B^{*-1}\tilde{S}_B)\right],
\end{aligned}
\tag{3.11}
$$

where the terms with the sign "*" imply that they depend on θ^*, those terms without the sign "*" imply that they are independent of θ^*, and

$$
\begin{aligned}
\tilde{\Sigma}_B^* &= \begin{pmatrix} \Sigma_B^* + \mu^*\mu^{*\prime} & \mu^* \\ \mu^{*\prime} & 1 \end{pmatrix}, \quad \tilde{S}_B = \begin{pmatrix} S_B & \bar{a} \\ \bar{a}' & 1 \end{pmatrix}, \\
\bar{a} &= \frac{1}{G}\sum_{g=1}^{G} a_g, \quad S_B = \frac{1}{G}\sum_{g=1}^{G} C_g, \\
S_{gW} &= \frac{1}{N_g}X_g'X_g - \bar{x}_g a_g' - a_g\bar{x}_g' + C_g.
\end{aligned}
\tag{3.12}
$$

When there is only one within covariance matrix, that is, $\Sigma_{gW} \equiv \Sigma_W$ for $g = 1,\ldots,G$, computation of the terms in Equation 3.12 can be substantially simplified. Let

$$
\Sigma_W = CC' \quad \text{and} \quad C^{-1}\Sigma_B(C^{-1})' = E\Delta E'
\tag{3.13}
$$

be the Cholesky decomposition of Σ_W and the eigenvector-eigenvalue decomposition of $C^{-1}\Sigma_B(C^{-1})'$, respectively, where C is a lower triangular matrix, E is an orthogonal matrix, and

$$
\Delta = \text{diag}(\delta_1,\ldots,\delta_p),
\tag{3.14}
$$

is a diagonal matrix whose diagonal elements can be arranged as $\delta_1 \geq \cdots \geq \delta_p > 0$ without loss of generality. By Equations 3.13 and 3.14, the inverse matrix $N_g \Sigma_g^{-1}$ in Equations 3.9 and 3.10 can be expressed as

$$N_g \Sigma_g^{-1} = N_g (\Sigma_W + N_g \Sigma_B)^{-1} = T' D_g T, \qquad (3.15)$$

with

$$T = (CE)^{-1}, \quad D_g = \mathrm{diag}\left(\frac{N_g}{1 + N_g \delta_1}, \cdots, \frac{N_g}{1 + N_g \delta_p}\right). \qquad (3.16)$$

Let

$$
\begin{aligned}
d_g &= D_g T(\bar{x}_g - \mu), \quad \bar{a}_0 = \frac{1}{G}\sum_{g=1}^{G} a_{g0}, \\
B &= \left\{\sum_{g=1}^{G}\frac{N_g}{N}(\bar{x}_g - \mu)d_g'\right\}(T\Sigma_B),
\end{aligned}
\qquad (3.17)
$$

where a_{g0} is given by Equation 3.10. Then in the case $\Sigma_{gW} \equiv \Sigma_W$, the E-step function Equation 3.11 reduces to

$$
\begin{aligned}
M(\theta^*|\theta) = \ &N\Big[\log|\Sigma_W^*| + \mathrm{tr}(\Sigma_W^{*-1} S_W)\Big] \\
&+G\Big[\log|\widetilde{\Sigma}_B^*| + \mathrm{tr}(\widetilde{\Sigma}_B^{*-1}\widetilde{S}_B)\Big],
\end{aligned}
\qquad (3.18)
$$

and the terms in Equation 3.12 reduce to

$$
\begin{aligned}
\bar{a} &= \mu + \bar{a}_0, \qquad \bar{a}_0 = (T\Sigma_B)'\Big(\frac{1}{G}\sum_{g=1}^{G} d_g\Big), \\
S_B &= \mu\mu' + \bar{a}_0\mu' + \mu\bar{a}_0' + \Sigma_B \\
&\quad +(T\Sigma_B)'\Big(\frac{1}{G}\sum_{g=1}^{G} d_g d_g' - \frac{1}{G}\sum_{g=1}^{G} D_g\Big)(T\Sigma_B) \\
S_W &= \frac{1}{N}\sum_{g=1}^{G} N_g S_{gW} \\
&= \frac{1}{N}X'X - B - B' + \mu\mu' - \bar{x}\mu' - \mu\bar{x}' + \Sigma_B \\
&\quad +(T\Sigma_B)'\Big(\sum_{g=1}^{G}\frac{N_g}{N} d_g d_g' - \sum_{g=1}^{G}\frac{N_g}{N} D_g\Big)(T\Sigma_B),
\end{aligned}
\qquad (3.19)
$$

and $\bar{x}$ is the overall sample mean as in Equation 3.2.

The M-step

By the principle of the EM algorithm, the M-step is to minimize the E-step function $M(\theta^*|\theta)$ given by Equation 3.11 or Equation 3.18 (for the case $\Sigma_{gW} \equiv \Sigma_W$) with regard to θ^* for each fixed θ. As pointed out by Lee and Poon (1998), Lange's (1995a) EM gradient algorithm is a convenient way to realize the M-step that converges to a local maximum of the ML function fast. The idea of the EM gradient algorithm is that for each $\theta = \theta_i$ at the i-th iteration (θ_0 corresponds to the starting value of θ), it is only necessary to find a value of $\theta^* = \theta_{i+1}$, such that

$$M(\theta_{i+1}|\theta_i) \le M(\theta_i|\theta_i). \tag{3.20}$$

This can be achieved by using the gradient direction of $M(\theta^*|\theta)$ at each fixed θ. From the simple E-step function Equation 3.11, it is easy to obtain:

$$
\begin{aligned}
d^{10}M(\theta|\theta) &\overset{\text{def}}{=} \frac{\partial}{\partial \theta^*} M(\theta^*|\theta)\Big|_{\theta^*=\theta} \\
&= \sum_{g=1}^{G} N_g \Delta_{gW} (\Sigma_{gW}^{-1} \otimes \Sigma_{gW}^{-1}) \text{vec}(\Sigma_{gW} - S_{gW}) \\
&\quad + G \, \widetilde{\Delta}_B \, (\widetilde{\Sigma}_B^{-1} \otimes \widetilde{\Sigma}_B^{-1}) \text{vec}(\widetilde{\Sigma}_B - \widetilde{S}_B),
\end{aligned}
\tag{3.21}
$$

where the sign "vec" denotes the vectorization of a matrix by stacking its columns successively, "$\otimes$" denotes the Kronecker product of matrices, $\widetilde{\Sigma}_B$ is defined in Equation 3.12 with $\theta^* = \theta$, and the matrices of derivatives

$$\Delta_{gW} = \frac{\partial(\text{vec}\Sigma_{gW})'}{\partial\theta}, \qquad \widetilde{\Delta}_B = \frac{\partial(\text{vec}\,\widetilde{\Sigma}_B)'}{\partial\theta}. \tag{3.22}$$

In the case $\Sigma_{gW} \equiv \Sigma_W$ ($\Delta_{gW} \equiv \Delta_W$), Equation 3.21 reduces to

$$
\begin{aligned}
d^{10}M(\theta|\theta) = \;& N\Delta_W(\Sigma_W^{-1} \otimes \Sigma_W^{-1})\text{vec}(\Sigma_W - S_W) \\
&+ G\, \widetilde{\Delta}_B \, (\widetilde{\Sigma}_B^{-1} \otimes \widetilde{\Sigma}_B^{-1})\text{vec}(\widetilde{\Sigma}_B - \widetilde{S}_B).
\end{aligned}
\tag{3.23}
$$

The Fisher information matrix can also be easily obtained:

$$
\begin{aligned}
I(\theta) &\overset{\text{def}}{=} \left\{ E\left[\frac{\partial^2}{\partial\theta^*\partial\theta^{*\prime}} M(\theta^*|\theta) \right] \right\}\Big|_{\theta^*=\theta} \\
&= \sum_{g=1}^{G} N_g \Delta_{gW} (\Sigma_{gW}^{-1} \otimes \Sigma_{gW}^{-1})\Delta_{gW}' \\
&\quad + G\, \widetilde{\Delta}_B \, (\widetilde{\Sigma}_B^{-1} \otimes \widetilde{\Sigma}_B^{-1})\, \widetilde{\Delta}_B'.
\end{aligned}
\tag{3.24}
$$

In the case $\Sigma_{gW} \equiv \Sigma_W$, Equation 3.24 reduces to

$$
\begin{aligned}
I(\theta) = & \; N\Delta_W (\Sigma_W^{-1} \otimes \Sigma_W^{-1})\Delta_W' \\
& + G\,\tilde{\Delta}_B\,(\tilde{\Sigma}_B^{-1} \otimes \tilde{\Sigma}_B^{-1})\,\tilde{\Delta}_B'.
\end{aligned}
\tag{3.25}
$$

The updating process of the EM gradient algorithm for obtaining the MLE of θ is given by

$$
\theta_{i+1} = \theta_i - \alpha I(\theta_i)^{-1} d^{10} M(\theta_i | \theta_i),
\tag{3.26}
$$

where $0 < \alpha \le 1$ is an adjusting constant, and θ_i denotes the value of θ at the i-th iteration.

SECTION 3: STANDARD ERROR AND TEST OF MODEL FIT

Asymptotic properties of the MLE of the model parameter vector θ from Equation 3.1 with zero mean structure (i.e., $\mu = 0$) were studied by Lee (1990). In the case that Equation 3.1 has both mean and covariance structures (Equation 3.3), approximate standard errors of model parameters can be obtained from the asymptotic normality of the MLE of θ. Hoadley (1971) gave some general results on the asymptotic normality of MLE for the case of independent nonidentically distributed (i.n.i.d. for simplicity) samples with some mild regularity conditions imposed on the underlying distribution. The MLE of θ, say, $\hat{\theta}$, minimizes the $-2 \log ML$ function, say, $f(\theta)$, of the i.n.i.d. sample $\{x_g : g = 1, \dots, G\}$ defined in Equation 3.5. Under the assumptions 1 through 4 in Section 1, $f(\theta)$ is given by

$$
\begin{aligned}
f(\theta) = & \sum_{g=1}^{G} (N_g - 1) \left\{ \log |\Sigma_{gW}| + \mathrm{tr}(\Sigma_{gW}^{-1} S_g) \right\} \\
& + \sum_{g=1}^{G} \left\{ \log |\Sigma_g| + \mathrm{tr}(\Sigma_g^{-1} U_g) \right\} + \sum_{g=1}^{G} N_g \mathrm{tr}[\Sigma_g^{-1} V_g(\mu)],
\end{aligned}
\tag{3.27}
$$

where

$$
S_g = \frac{1}{N_g - 1} \sum_{i=1}^{N_g} (x_{gi} - \bar{x}_g)(x_{gi} - \bar{x}_g)',
\tag{3.28}
$$

$$
U_g = N_g \bar{x}_g \bar{x}_g', \quad V_g(\mu) = \mu\mu' - \bar{x}_g\mu' - \mu\bar{x}_g'.
$$

It can be verified that $f(\theta)$ satisfies the regularity conditions in *Theorem 2* of Hoadley (1971). Thus, we have

Theorem 1. The MLE $\hat{\boldsymbol{\theta}}$ that minimizes $f(\boldsymbol{\theta})$ given by Equation 3.27 has the asymptotic normal distribution

$$G^{1/2}(\hat{\boldsymbol{\theta}} - \boldsymbol{\theta}) \xrightarrow{\mathcal{D}} N(0, 2\boldsymbol{\Gamma}^{-1}(\boldsymbol{\theta})), \quad G \to \infty, \tag{3.29}$$

where the sign "$\xrightarrow{\mathcal{D}}$" means "converge in distribution," and the matrix $\boldsymbol{\Gamma}(\boldsymbol{\theta})$ is approximately given by

$$
\begin{aligned}
\boldsymbol{\Gamma}(\boldsymbol{\theta}) &= \frac{1}{G} \cdot E\left\{ \frac{\partial^2 f(\boldsymbol{\theta})}{\partial \boldsymbol{\theta} \partial \boldsymbol{\theta}'} \right\} \\
&= \frac{1}{G} \sum_{g=1}^{G} \left\{ (N_g - 1)\boldsymbol{\Delta}_{gW}(\boldsymbol{\Sigma}_{gW}^{-1} \otimes \boldsymbol{\Sigma}_{gW}^{-1})\boldsymbol{\Delta}_{gW}' \right. \\
&\quad \left. + \boldsymbol{\Delta}_g(\boldsymbol{\Sigma}_g^{-1} \otimes \boldsymbol{\Sigma}_g^{-1})\boldsymbol{\Delta}_g' + 2N_g\boldsymbol{\Delta}_\mu\boldsymbol{\Sigma}_g^{-1}\boldsymbol{\Delta}_\mu' \right\},
\end{aligned}
\tag{3.30}
$$

where the terms $\boldsymbol{\Delta}_{gW}$ (defined in Equation 3.22), $\boldsymbol{\Delta}_B$, and $\boldsymbol{\Delta}_\mu$ are matrices of derivatives, and

$$\boldsymbol{\Delta}_B = \frac{\partial(\mathrm{vec}\boldsymbol{\Sigma}_B)'}{\partial \boldsymbol{\theta}}, \quad \boldsymbol{\Delta}_g = \frac{\partial(\mathrm{vec}\boldsymbol{\Sigma}_g)'}{\partial \boldsymbol{\theta}} = \boldsymbol{\Delta}_{gW} + N_g\boldsymbol{\Delta}_B, \quad \boldsymbol{\Delta}_\mu = \frac{\partial \boldsymbol{\mu}'}{\partial \boldsymbol{\theta}}. \tag{3.31}$$

By Theorem 1, the asymptotic standard errors of the components of the MLE $\hat{\boldsymbol{\theta}}$ can be approximately computed from the corresponding diagonal components of the matrix

$$\frac{2}{G}\boldsymbol{\Gamma}(\hat{\boldsymbol{\theta}})^{-1}. \tag{3.32}$$

In the case $\boldsymbol{\Sigma}_{gW} \equiv \boldsymbol{\Sigma}_W$, a simplified formula for computing matrix (Equation 3.30) can be provided by using the decompositions in Equations 3.13 and 3.15:

$$
\begin{aligned}
\boldsymbol{\Gamma}(\boldsymbol{\theta}) =\ & (\frac{N}{G} - 1)\boldsymbol{\Delta}_W(\boldsymbol{\Sigma}_W^{-1} \otimes \boldsymbol{\Sigma}_W^{-1})\boldsymbol{\Delta}_W' \\
&+ \boldsymbol{\Delta}_W(\boldsymbol{T} \otimes \boldsymbol{T})\boldsymbol{D}^{(1)}[\boldsymbol{\Delta}_W(\boldsymbol{T} \otimes \boldsymbol{T})]' \\
&+ \boldsymbol{\Delta}_W(\boldsymbol{T} \otimes \boldsymbol{T})\boldsymbol{D}^{(2)}[\boldsymbol{\Delta}_B(\boldsymbol{T} \otimes \boldsymbol{T})]' \\
&+ \{\boldsymbol{\Delta}_W(\boldsymbol{T} \otimes \boldsymbol{T})\boldsymbol{D}^{(2)}[\boldsymbol{\Delta}_B(\boldsymbol{T} \otimes \boldsymbol{T})]'\}' \\
&+ \boldsymbol{\Delta}_B(\boldsymbol{T} \otimes \boldsymbol{T})\boldsymbol{D}^{(3)}[\boldsymbol{\Delta}_B(\boldsymbol{T} \otimes \boldsymbol{T})]' \\
&+ 2(\boldsymbol{\Delta}_\mu\boldsymbol{T})\bar{\boldsymbol{D}}(\boldsymbol{\Delta}_\mu\boldsymbol{T})',
\end{aligned}
\tag{3.33}
$$

where

$$\boldsymbol{D}^{(1)} = \frac{1}{G}\sum_{g=1}^{G} N_g^{-2}\boldsymbol{D}_g \otimes \boldsymbol{D}_g, \qquad \boldsymbol{D}^{(3)} = \frac{1}{G}\sum_{g=1}^{G} \boldsymbol{D}_g \otimes \boldsymbol{D}_g,$$

$$\boldsymbol{D}^{(2)} = \frac{1}{G}\sum_{g=1}^{G} N_g^{-1}\boldsymbol{D}_g \otimes \boldsymbol{D}_g, \qquad \bar{\boldsymbol{D}} = \frac{1}{G}\sum_{g=1}^{G} \boldsymbol{D}_g.$$

(3.34)

Then $\boldsymbol{\Gamma}(\hat{\boldsymbol{\theta}})$ can be computed by substituting $\hat{\boldsymbol{\theta}}$ for $\boldsymbol{\theta}$ in Equation 3.33.

It can be verified that $\boldsymbol{D}^{(1)}$, $\boldsymbol{D}^{(2)}$, and $\boldsymbol{D}^{(3)}$ are still diagonal matrices with their diagonal elements given by

$$\boldsymbol{D}^{(1)}(k) = \frac{1}{G}\sum_{g=1}^{G} N_g^{-2} d_{g,[(k-1)/p]+1} d_{g,k-p[(k-1)/p]},$$

$$\boldsymbol{D}^{(2)}(k) = \frac{1}{G}\sum_{g=1}^{G} N_g^{-1} d_{g,[(k-1)/p]+1} d_{g,k-p[(k-1)/p]},$$

(3.35)

$$\boldsymbol{D}^{(3)}(k) = \frac{1}{G}\sum_{g=1}^{G} d_{g,[(k-1)/p]+1} d_{g,k-p[(k-1)/p]},$$

for $k = 1,\ldots,p^2$, where $d_{g,j} = N_g/(1 + N_g\delta_j)$ is the j-th diagonal component of $\boldsymbol{D}_g$ and the sign $[\cdot]$ denotes the integer part of a real number (e.g., $[2.9] = 2$ and $[3.1] = 3$).

A test of model fit can be constructed using standard ML theory. The asymptotic chi-squared statistic for testing model fit is given by

$$\chi_L^2 = f(\hat{\boldsymbol{\theta}}) - f(\hat{\boldsymbol{\theta}}_s) \overset{D}{\to} \chi^2(m) \quad (G \to \infty),$$

(3.36)

where $\hat{\boldsymbol{\theta}}$ denotes the MLE of $\boldsymbol{\theta}$ in the structured case Equation 3.3, (the null hypothesis), $\hat{\boldsymbol{\theta}}_s$ denotes the MLE of $\boldsymbol{\theta}_s$, which is the model parameter vector in the unstructured (or saturated) case (the alternative hypothesis):

$$\boldsymbol{\theta}_s = \left[(\text{vecs}\boldsymbol{\Sigma}_{1W})',\ldots,(\text{vecs}\boldsymbol{\Sigma}_{GW})',(\text{vecs}\boldsymbol{\Sigma}_B)',\boldsymbol{\mu}'\right]',$$

(3.37)

where the sign "vecs" operated on a symmetric matrix constitutes a column vector stacked by the nonduplicated elements in the symmetric matrix. In the saturated case, all nonduplicated elements in $\boldsymbol{\Sigma}_B$, $\boldsymbol{\Sigma}_{gW}$ $(g = 1,\ldots,G)$ and the components in $\boldsymbol{\mu} = (\mu_1,\ldots,\mu_p)'$ are free model parameters. Therefore, the number of degrees of freedom (df) m in Equation 3.36 is given by

$$m = \begin{cases} (G+1)p(p+1)/2 + p - r, & \text{if } \boldsymbol{\Sigma}_{gW}\text{'s are mutually distinguished,} \\ p^2 + 2p - r, & \text{if } \boldsymbol{\Sigma}_{gW} \equiv \boldsymbol{\Sigma}_W, \end{cases}$$

(3.38)

where r is the number of free parameters in the structured case (Equation 3.3).

Computation of $\hat{\boldsymbol{\theta}}_s$ for the saturated model can be carried out by the following iteration process. In the E-step function Equation 3.11 or 3.18, we can consider $\boldsymbol{\theta}^*$ and $\boldsymbol{\theta}$ as any two specified values of $\boldsymbol{\theta}_s$ because either Equation is true for any structured case specified by Equation 3.3 and when the parameter $\boldsymbol{\theta}$ in Equation 3.3 takes the form of $\boldsymbol{\theta}_s$, the model defined by Equation 3.1 becomes saturated. At the i-th iteration ($i = 0$ corresponds to the case of starting values), let $\boldsymbol{S}_{gW}(i)$, $\boldsymbol{S}_B(i)$, $\boldsymbol{\mu}(i)$, and $\bar{\boldsymbol{a}}(i)$ be the values of $\boldsymbol{S}_{gW}$, $\boldsymbol{S}_B$, $\boldsymbol{\mu}$, and $\bar{\boldsymbol{a}}$, respectively; at the $(i + 1)$-th iteration, let $\boldsymbol{S}_{gW}(i+1)$, $\boldsymbol{S}_B(i+1)$, $\boldsymbol{\mu}(i+1)$, and $\bar{\boldsymbol{a}}(i+1)$ be the values of $\boldsymbol{S}_{gW}$, $\boldsymbol{S}_B$, $\boldsymbol{\mu}$, and $\bar{\boldsymbol{a}}$, respectively. Because $\boldsymbol{S}_{gW}(i+1)$, $\boldsymbol{S}_B(i+1)$, $\boldsymbol{\mu}(i+1)$, and $\bar{\boldsymbol{a}}(i+1)$ are supposed to minimize the E-step function in equation 11 at the $(i + 1)$-th iteration in the saturated case, we can obtain

$$
\begin{aligned}
\boldsymbol{\Sigma}_{gW}(i + 1) &= \boldsymbol{S}_{gW}(i), \\
\boldsymbol{\Sigma}_B(i + 1) &= \boldsymbol{S}_B(i) - \bar{\boldsymbol{a}}(i)\bar{\boldsymbol{a}}(i)', \\
\boldsymbol{\mu}(i + 1) &= \bar{\boldsymbol{a}}(i).
\end{aligned}
\tag{3.39}
$$

The updating process Equation 3.39 can stop when some given convergence criterion is reached.

In the case $\boldsymbol{\Sigma}_{gW} \equiv \boldsymbol{\Sigma}_W$, a simplified formula for computing $f(\boldsymbol{\theta})$ given by Equation 3.27 can also be provided by using the decompositions in Equations 3.13 and 3.15. It turns out that

$$
f(\boldsymbol{\theta}) = N \log|\boldsymbol{\Sigma}_W| + \mathrm{tr}[\boldsymbol{\Sigma}_W^{-1}\, \tilde{\boldsymbol{S}}_W] + \sum_{g=1}^{G}\sum_{i=1}^{p} \log(1 + N_g\delta_i) + \sum_{g=1}^{G}\sum_{i=1}^{p} \frac{N_g c_{gi}^2}{1 + N_g\delta_i},
\tag{3.40}
$$

where the δ_i's are given in Equation 3.14, and

$$
\tilde{\boldsymbol{S}}_W = \sum_{g=1}^{G} \boldsymbol{X}_g'\left(\boldsymbol{I}_{N_g} - \frac{1}{N_g}\mathbf{1}_{N_g}\mathbf{1}_{N_g}'\right)\boldsymbol{X}_g,
\tag{3.41}
$$

$$
(c_{g1},\dots,c_{gp})' \stackrel{\text{def}}{=} \boldsymbol{T}(\bar{\boldsymbol{x}}_g - \boldsymbol{\mu}),
$$

where $\boldsymbol{\mu} = \boldsymbol{\mu}(\boldsymbol{\theta})$ given by Equation 3.3 in computing $f(\hat{\boldsymbol{\theta}})$ for the structured case and $\boldsymbol{\mu} = (\mu_1,\dots,\mu_p)'$ in computing $f(\hat{\boldsymbol{\theta}}_s)$ for the saturated case. $\boldsymbol{I}_{N_g}$ is an $N_g \times N_g$ identity matrix, $\mathbf{1}_{N_g}$ is n $N_g \times 1$ vector of ones, and $\boldsymbol{X}_g$ and $\boldsymbol{T}$ are defined in Equations 3.6 and 3.16, respectively.

SECTION 4: AN EXAMPLE

The two-level "Alcohol Use Data" were collected and analyzed by Duncan et al. (1998) by using the technique of longitudinal latent variable modeling. The data are from responses regarding use of four types of alcohol by siblings (individuals or level-1 units) of at least 11 years of age in $G = 435$ families (groups or level-2 units). That is, we have $p = 4$ observable variables. There are five distinct cluster (family or group) sizes (i.e., the N_g, $g = 1, \ldots, G$) ranging from 2 to 6 members. The total sample size (total number of individuals) is $N = \sum_{g=1}^{G} N_g = 1204$. The complete data set is stored in a data file "duna.dat" in M*plus* (Muthén & Muthén, 1998) examples website: (http://www.statmodel.com/mplus/examples/). In this section, we carry out the ML analysis on this data set by using the EM gradient algorithm developed in Section 2. Standard errors of the estimates of model parameters and the chi-squared test of model fit derived from Section 3 are also provided. In order to compare the results of our ML analysis with those provided by M*plus*, we set up the same model with a mean structure and ran M*plus* for the same data set. A two-level confirmatory factor analysis model is set up for both the ML analysis and the MUML analysis. We consider the case $\Sigma_{gW} \equiv \Sigma_W$. Because the algorithm in Section 2 has been coded in EQS (Bentler, 2002), we just present the model by EQS commands:

$$
\begin{array}{lll}
\text{within:} & V_1 = 1F1 + 0F2 + E1, & \text{between:} \quad V_1 = 1F1 + 0F2 + E1, \\
& V_2 = 1F1 + 1F2 + E2, & \qquad\qquad V_2 = 1F1 + 1F2 + E2, \\
& V_3 = 1F1 + *F2 + E3, & \qquad\qquad V_3 = 1F1 + *F2 + E3, \\
& V_4 = 1F1 + *F2 + E4, & \qquad\qquad V_4 = 1F1 + *F2 + E4.
\end{array}
$$

$$(3.42)$$

This implies that both the within and the between models have the same structure. They are confirmatory factor analysis models with two factors. In Equation 3.42, the sign "$*$" before the factor $F2$ stands for an unknown factor loading parameter, and the constants "1" and "0" before $F1$ and $F2$ for fixed factor loadings, $E1$, $E2$, $E3$, $E4$ are uncorrelated residual variables with zero means and unknown variances. The assumptions 1 through 4 in Section 1 are imposed on Equation 3.42. This equation implies that the within and between covariance matrices are given by

$$
\Sigma_W = \Lambda_W \Phi_W \Lambda_W' + \Psi_W, \quad \Sigma_B = \Lambda_B \Phi_B \Lambda_B' + \Psi_B \tag{3.43}
$$

where

$$
\Lambda_W' = \begin{pmatrix} 1 & 1 & 1 & 1 \\ 0 & 1 & \theta_1 & \theta_2 \end{pmatrix}, \quad \Lambda_B' = \begin{pmatrix} 1 & 1 & 1 & 1 \\ 0 & 1 & \phi_1 & \phi_2 \end{pmatrix}, \tag{3.44}
$$

and

$$\mathbf{\Phi}_W = \begin{pmatrix} \theta_3 & \theta_5 \\ \theta_5 & \theta_4 \end{pmatrix}, \qquad \mathbf{\Phi}_B = \begin{pmatrix} \phi_3 & \phi_5 \\ \phi_5 & \phi_4 \end{pmatrix},$$

$$\mathbf{\Psi}_W = \text{diag}(\theta_6, \theta_7, \theta_8, \theta_9), \qquad \mathbf{\Phi}_B = \text{diag}(\phi_6, \phi_7, \phi_8, \phi_9). \tag{3.45}$$

Assume that the two factors $F1$ and $F2$ in the between model in Equation 3.42 have unknown nonzero means:

$$E(F1) = \mu_1, \qquad E(F2) = \mu_2, \tag{3.46}$$

while the two factors in the within model in Equation 3.42 have zero means. Then we can consider that Equation 3.42 has a mean structure

$$\mu = \mathbf{\Lambda}_B (\mu_1, \mu_2)' = (\mu_1, \mu_1 + \mu_2, \mu_1 + \phi_1 \mu_2, \mu_1 + \phi_2 \mu_2)'. \tag{3.47}$$

Based on this mean structure and the covariance structure specified by Equations 3.43 through 3.45, we can easily obtain the matrices of derivatives $\mathbf{\Delta}_W$ and $\widetilde{\mathbf{\Delta}}_B$ in implementing the EM gradient algorithm in Section 2. Starting values for the model parameters θ_i, ϕ_i $(i = 1, \ldots, 9)$, μ_1, and μ_2 need to be given in carrying out the iteration process defined by Equation 3.26. Theoretically, the convergence of the EM algorithm does not depend on the choice of starting values of model parameters. A good choice of a set of starting values may result in fast convergence of an EM algorithm. Because we do not know the possible true values of the model parameters, we choose a set of randomly generated starting values: (a) θ_1 and θ_2 are chosen as two random numbers in $(0, 1)$; (b) θ_3 and θ_4 as two random numbers in $(0.5, 1.5)$; (c) θ_5 as a random number in $(0, 0.5)$; (d) θ_6 to θ_9 as random numbers in $(0.2, 1.2)$; (e) μ_1 and μ_2 as random numbers in $(0, 2)$. The starting values of ϕ_1 through ϕ_9 are chosen in exactly the same way as for θ_1 through θ_9. In computing the MLE of $\boldsymbol{\theta}_s = \{\text{vecs}\mathbf{\Sigma}_W, \text{vecs}\mathbf{\Sigma}_B, \boldsymbol{\mu}\}$ $[\boldsymbol{\mu} = (\mu_1, \mu_2, \mu_3, \mu_4)']$ for the saturated model, we choose the starting values for $\mathbf{\Sigma}_W$, $\mathbf{\Sigma}_B$, and $\boldsymbol{\mu}$ as $\mathbf{\Sigma}_W = \mathbf{\Sigma}_B = \widetilde{\mathbf{S}}_W / (N - G)$ and $\boldsymbol{\mu} = \bar{x}$, where $\widetilde{\mathbf{S}}_W$ is defined in Equation 3.41, and $\bar{x}$ is the overall sample mean from the data. In principle, the constant α in controlling the step length in the iteration process Equation 3.26 can be selected dynamically. That is, in each iteration, α could be chosen as the largest value $(0 < \alpha \le 1)$ that would reduce Equation 3.20 and keep all covariance matrices positive definite. In this example, we tried $\alpha \equiv 1$ in all iterations and it turns out that the algorithm is convergent in the sense that the *Root Mean Square Error* (RMSE for simplicity) converges to zero. The RMSE was used by Lee and Poon (1998) as a criterion for convergence of the EM algorithm.

Let $\boldsymbol{\theta}_i$ $(r \times 1)$ denote the value of $\boldsymbol{\theta}$ $(r \times 1)$ at the i-th iteration. The RMSE between two consecutive iterations is defined by

$$RMSE(\boldsymbol{\theta}) = \left\{ \frac{1}{r} \|\boldsymbol{\theta}_{i+1} - \boldsymbol{\theta}_i\|^2 \right\}^{1/2}, \qquad (3.48)$$

where the sign $\| \cdot \|$ stands for the Euclidean norm of a vector. For the data set duna.dat in Equation 3.42, we choose the convergence criterion for the EM gradient algorithm as $RMSE \leq 10^{-6}$ (the same as for M*plus*). The algorithm converges after 246 iterations and it took approximately $1\frac{1}{2}$ minutes of CPU time on a *Pentium II* PC to obtain the ML results. The RMSE from the updating process Equation 3.26 is actually already less than 10^{-4} after 45 iterations. This implies that the algorithm converges very fast in the sense that the RMSE cannot be improved much after 45 iterations. The ML results and those provided by running M*plus* for Equation 3.42 with fixed intercepts of 0 for the between model and unknown means for the two factors are presented in Table 3.1. The standard errors for the MLE of model parameters (by the ML analysis) are computed from the square roots of the asymptotic covariance matrix given by Equation 3.32. The chi-square is computed by Equation 3.36 and the degress of freedom for the ML analysis is computed by Equation 3.38. From Table 3.1, we can see that all ML results are close to the corresponding MUML results. But the standard errors provided by the ML analysis are generally a little bit smaller than those corresponding standard errors provided by MUML analysis. For this data set with an unbalanced design, the cluster sizes range from 2 to 6. M*plus* carries out the MUML analysis by using the average cluster size $s = (N^2 - \sum_{g=1}^{G} N_g^2)/[N(G-1)] = 2.767$ as the common cluster size. According to Muthén (1991, 1994), Hox (1993), and McDonald (1994), the pseudobalanced estimates provided by MUML analysis usually give good approximation to the full ML estimates. Our full ML analysis on the "Alcohol Use Data" is consistent with this observation.

CONCLUDING REMARKS

The mean and covariance structure model considered in this chapter is a small extension to the covariance structure model studied by Lee and Poon (1998). The EM gradient algorithm in this chapter was developed with the same technique as that used by Lee and Poon (1998). In the case of only one within covariance matrix, the utilization of the Cholesky decomposition and the eigenvector–eigenvalue decomposition in simplifying the computation of the E-step function in Section 2 greatly reduces the computational burden by avoiding the computation of a large number of inverse matrices. Therefore, the algorithm developed in this chapter is still

TABLE 3.1

ML and MUML Estimates, Standard Errors, and Tests

	θ_1	θ_2	θ_3	θ_4	θ_5	θ_6	θ_7
ML	1.722	1.874	0.663	0.111	-.011	0.188	0.244
MUML	1.739	1.868	0.663	0.108	-.011	0.187	0.247
S.E.(ML)	0.180	0.215	0.053	0.034	0.028	0.042	0.016
S.E.(MUML)	0.213	0.245	0.053	0.038	0.031	0.052	0.024
	θ_8	θ_9	ϕ_1	ϕ_2	ϕ_3	ϕ_4	ϕ_5
ML	0.122	0.356	1.767	3.624	0.285	0.004	-.035
MUML	0.117	0.361	1.755	3.575	0.287	0.005	-.035
S.E.(ML)	0.022	0.031	0.393	0.912	0.046	0.005	0.014
S.E.(MUML)	0.029	0.042	0.415	1.030	0.044	0.006	0.015
	ϕ_6	ϕ_7	ϕ_8	ϕ_9	μ_1	μ_2	
ML	0.055	0.022	0.043	0.073	1.978	0.079	
MUML	0.058	0.020	0.046	0.067	2.003	0.076	
S.E.(ML)	0.020	0.012	0.012	0.028	0.039	0.024	
S.E.(MUML)	0.026	0.014	0.016	0.037	0.038	0.026	
ML	chi-square= 6.311		df= 4	p-value= 0.1771			
MUML	chi-square= 6.237		df= 4	p-value= 0.1817			

Note. In the Table 3.1, S.E.(ML) stands for standard error from ML analysis; S.E.(MUML) stands for standard error from MUML analysis.

an improvement over that of Lee and Poon (1998), even in the case of the only covariance structure ($\mu = 0$) for Equation 3.1. It can be easily verified that the E-step function (Equation 3.11 or 3.18) reduces to that derived by Lee and Poon (1998) when no mean structure is involved in Equation 3.1.

As pointed out in Section 1, Equation 3.1 covers several interesting mean and covariance structure models. In ML analysis of two-level factor analysis models, it is meaningful to hypothesize that the factors have unknown nonzero means. Such models can be formulated as Equation 3.1 with both mean and covariance structures. Ignorance of such a mean structure may result in poor model fit by the chi-square test and less accurate (larger standard error) estimates (Yung & Bentler, 1999). The models considered in Muthén (1994) and some latent growth models studied later by Muthén (1997) can also be formulated as Equation 3.1 with both mean and covariance structures when some latent variables have unknown nonzero means. Therefore, the method for ML analysis of two-level SEM developed in this chapter is applicable to a wide range of two-level latent variable models. A noticeable feature of the algorithm in this chapter is that it avoids using the average group sample size to approximate the diverse group sample sizes in unbalanced sample designs that was suggested by Muthén (1994, 1997).

Finally, we point out the fact that the EM gradient algorithm is at least locally convergent in the sense that the series $\{\theta_i\}$ generated from the M-step Equation 3.26 converges to a stationery point. This point must give a local minimum of the $-2\log ML$ function (Equation 3.27) because the E-step function (Equation 3.11 or 3.18) always decreases along its gradient direction. The local convergence of general EM gradient algorithm was also discussed by Lange (1995a). Boyles (1983) and Wu (1983) studied some general conditions on the convergence of EM algorithms. It is noted that the EM gradient algorithm is not the only one to minimize the E-step function (Equation 3.11 or 3.18). Some existing accelerating EM algorithms, such as those proposed by Lange (1995b) and Jamshidian and Jennrich (1997), could also be employed. Further, the constant α for controlling the step length in the M-step defined by Equation 3.26 could be chosen dynamically at each iteration to improve the convergence of the EM gradient algorithm. This means that $0 < \alpha \leq 1$ can be chosen as large as possible while ensuring Equation 3.20 and keeping the estimated covariance matrices positive definite at each iteration.

ACKNOWLEDGMENT

This work was supported by National Institute on Drug Abuse grants DA01070 and DA00017.

REFERNCES

Bentler, P. M. (2002). *EQS 6 structural equations program manual.* Encino, CA: Multivariate Software, Inc.

Boyles, R. A. (1983). On the convergence of the EM algorithm. *Journal of the Royal Statistical Society, 45* (Series B), 47–50.

Dempster, A. P., Laird, N. M., & Rubin, D. B. (1977). Maximum likelihood from incomplete data via EM algorithm (with discussion). *Journal of the Royal Statistical Society, 39* (Series B), 1–38.

Duncan, T. E., Duncan, S. C., Alpert, A., ops, H., Stoolmiller, M., & Muthén, B. O. (1998). Latent variable modeling of longitudinal and multilevel substance use data. *Multivariate Behavioral Research, 32,* 275–318.

Goldstein, H., & McDonald, R. P. (1988). A general model for the analysis of multilevel data. *Psychometrika, 53,* 455–467.

Heck, R. H., & Thomas, S. L. (2000). *An introduction to multilevel modeling techniques.* Mahwah, NJ: Lawrence Erlbaum Associates.

Hoadley, B. (1971). Asymptotic properties of maximum likelihood estimators for the independent not identically distributed case. *Annals of Mathematical Statistics, 42,* 1977–1991.

Hox, J. J. (1993). Factor analysis of multilevel data: Gauging the Muthén model. In J. H. L. Oud, & R. A. W. van Blokland-Vogelesang (Eds.), *Advances in longitudinal and multivariate analysis in the behavioral sciences* (pp. 141–156). Nijmegen, The Netherlands: Instituur vor Toegepaste Sociale Wetenschappen.

Hox, J. J. (2000). Multilevel analyses of grouped and longitudinal data. In T. D. Little, K. U. Schnabel, & J. Baumert (Eds.), *Modeling longitudinal and multilevel data* (pp. 15–32). Mahwah, NJ: Lawrence Erlbaum Associates.

Jamshidian, M., & Jennrich, R. I. (1997). Acceleration of the EM algorithm by using Quasi-Newton methods. *Journal of the Royal Statistical Society, 59* (Series B), 569–587.

Lange, K. (1995a). A gradient algorithm locally equivalent to the EM algorithm. *Journal of the Royal Statistical Society, 57* (Series B), 425–437.

Lange, K. (1995b). A Quasi-Newton acceleration of the EM algorithm. *Statistica Sinica, 5,* 1–18.

Lee, S. Y. (1990). Multilevel analysis of structural equation models. *Biometrika, 77,* 763–772.

Lee, S. Y., & Poon, W. Y. (1998). Analysis of two-level structural equation models via EM type algorithms. *Statistica Sinica, 8,* 749–766.

McDonald, R. P. (1994). The bilevel reticular action model for path analysis with latent variables. *Sociological Methods & Research, 22,* 399–413.

McDonald, R. P., & Goldstein, H. (1989). Balanced versus unbalanced designs for linear structural relations in two-level data. *British Journal of Mathematical & Statistical Psychology, 42,* 215–232.

Muthén, B. O. (1991). Multilevel factor analysis of class and student achievement components. *Journal of Educational Measurement, 28,* 338–354.

Muthén, B. O. (1994). Multilevel covariance structure analysis. *Sociological Methods & Research, 22,* 376–398.

Muthén, B. O. (1997). Latent variable modeling of longitudinal and multilevel data. In A. E. Raftery (Ed.), *Sociological methodology 1997* (pp. 453–481). Washington, DC: American Sociological Association.

Muthén, L. K., & Muthén, B. O. (1998). *Mplus user's guide*. Los Angeles: Muthén & Muthén, Inc.

Raudenbush, S. W. (1995). Maximum likelihood estimation for unbalanced multilevel covariance structure models via the EM algorithm. *British Journal of Mathematical & Statistical Psychology, 48*, 359–370.

Wu, C. F. J. (1983). On the convergence properties of the EM algorithm. *Annals of Statistics, 11*, 95–103.

Yung, Y. F., & Bentler, P. M. (1999). On added information for ML factor analysis with mean and covariance structures. *Journal of Educational and Behavioral Statistics, 24*, 1–20.

4

Analysis of Reading Skills Development from Kindergarten through First Grade: An Application Of Growth Mixture Modeling To Sequential Processes

Bengt Muthén
University of California, Los Angeles

Siek-Toon Khoo
Arizona State University

David J. Francis
University of Houston

Christy K. Boscardin
University of California, Los Angeles

This chapter outlines how growth mixture modeling can be used to study achievement and learning progress. The work is motivated by a study of reading development among children from kindergarten to first grade. Section 1 presents the data and the substantive problem. Section 2 discusses random coefficient growth modeling. Section 3 presents how random coefficient growth modeling in a latent variable framework can be used to relate the growth factors of two growth processes. Section 4 extends the latent variable framework so that multiple classes of development can be studied.

SECTION 1: THE SUBSTANTIVE PROBLEM

The reading study

The research questions originated from the study entitled "Detecting Reading Problems by Modeling Individual Growth" (Francis, 1996), also referred to as the EARS study (Early Assessment of Reading Skills). EARS collected data in a modified longitudinal time-sequential design involving about 1,000 children. The children were measured four times a year from kindergarten to grade two. In grades one and two, measures included spelling, word recognition, and reading comprehension. In kindergarten, skills that are considered precursor skills to reading development were measured, such as alphabetic awareness, orthographic and phonemic awareness, and visual motor integration. Standardized reading comprehension tests were administered at the end of first and second grade. The background variables gender, SES, and ethnicity were collected.

Francis (1996) focused on the early detection and identification of reading–disabled children. In this context, he formulated three research hypotheses; (1) kindergarten children will differ in their growth and development in precursor skills; (2) the rate of development of the precursor skills will relate to the rate of development and the level of attainment of reading and spelling skills, and individual growth rates in reading and spelling skills will predict performance on standardized tests of reading and spelling; (3) the use of growth rates for skills and precursors will allow for earlier identification of children at risk for poor academic outcomes and lead to more stable predictions regarding future academic performance.

General issues

Conventional growth modeling of individual differences in development can, in principle, use growth trajectory features such as the rate of learning as statistically based measures of progress. There is a general problem, however, of measuring and modeling student progress over an extended period of time. As the EARS study illustrates, the underlying construct under study in a developmental process is changing and evolving due to maturation of subjects. Reading skills are relevant in first grade but not in kindergarten. In kindergarten, reading precursor skills are of interest, but lose their relevance in first grade.

This exposes the Achilles heel of growth modeling, namely the assumption that the outcome variable has a constant scale or metric and a stable meaning over time. If it does not, conventional growth modeling is not meaningful. Item response theory offers a limited solution to this problem by allowing the formation of scale scores based on different test forms that change over time but that have overlapping items. But constructs of interest in a longitudinal study are naturally changing and evolving over time in more fundamental ways and to capture this, a more radical solution is necessary.

Changing meaning of the outcome does not make growth modeling impossible. Instead, conventional growth modeling needs to be developed methodologically to suit the research problem. Developmental processes that evolve over time need to be studied in the context of multistage growth and multiple processes. There is a need to investigate modeling methodology that can describe how one growth process leads into the next process. It is of interest to see how relationships between trajectories of early growth processes relate to failure or success in later growth processes.

The solution proposed in this chapter is essentially to turn the problem into an opportunity. Different developmental phases have different expressions of a construct and should not be forced onto the same scale. Instead, a multistage analysis approach should be taken where the different phases are viewed as sequential processes, one leading to another, and are analyzed jointly. This study focuses on how an early process influences a later process as exemplified by how the development of phonemic awareness during kindergarten influences the development of word recognition in first grade. Our special focus is on modeling that provides a prediction of first-grade development by kindergarten development.

SECTION 2: GROWTH MODELING

Research hypotheses regarding achievement and learning are often formulated in terms of individual development over time and tested using repeated measurements on groups of individuals. With a developmental perspective, the interest is not so much in the level of a certain outcome at a particular time point as it is in the growth trajectory across multiple time points. Learning outcomes typically show natural systematic growth over time. There may be an initial phase of rapid increase followed by a later phase of leveling out. The starting level, the rate of increase, and the leveling out are of interest in studying learning theories. The focus is on characterizing the individual variation in development and describing it in terms of its antecedents and consequences.

Standard statistical techniques for repeated measures data use random coefficient modeling to describe individual differences in development. This is carried out using software such as BMDP5V, SAS PROC MIXED, and MIXOR using the mixed linear model (Jennrich & Schluchter, 1986; Laird & Ware, 1982; Lindstrom & Bates, 1988), or MLn and HLM drawing on hierarchical linear (multilevel) modeling (Bryk & Raudenbush, 1992; Goldstein, 1995). From a modeling point of view, these approaches are essentially the same. Although it is possible to model multivariate outcomes using these techniques (see, MacCallum, Kim, Malarkey, & Kiecolt-Glaser, 1997; Thum, 1997), applications typically focus on longitudinal development of a univariate outcome variable. Antecedents of individual variation are modeled as time-invariant covariates whereas time-specific antecedents are modeled as time-varying covariates.

Developmental theories can be better modeled if the analysis methodology can allow trajectory shapes to be of primary focus rather than measurements at specific time points. This means that analysis methodology is needed to describe trajectory shapes not only as outcomes, but also as predictors, as mediators, and, in intervention studies, as the performance of a control group to which the trajectories of the intervention group are compared. Multiple processes, each with its own set of trajectories, for which the interplay and dependencies of the processes are of key interest should also be allowed. The trajectories should be able to have multiple indicators at each time point to reduce measurement error influence and to capture several aspects of the developing construct.

Given this broader research perspective, it is advantageous to perform repeated measures analysis in a more general framework than in the mixed linear model or multilevel model. Latent variable structural equation modeling offers such a general framework. Although repeated measures analysis of a single outcome variable is obtained as a special case of latent variable structural equation modeling, the generalizations discussed earlier are possible in the latent variable structural equation modeling framework. This is because the random coefficients are represented as latent variables where the latent variables can have regression relations among themselves and where the latent variables can also represent constructs as outcomes that have multiple indicators. Using psychometric growth modeling introduced by Meredith and Tisak (1990) as a starting point, Muthén and Curran (1997) gave an overview of latent variable work related to longitudinal modeling as well as mixed linear modeling and hierarchical linear modeling work and provided an up-to-date account of the potential of latent variable techniques for longitudinal data suitable for developmental studies. As pointed out in Muthén and Curran (1997), once the mixed linear model is put into the latent variable structural equation modeling framework, many general forms of longitudinal analysis are possible, including mediational variables influencing the developmental process; ultimate (distal) outcome variables influenced by the developmental process; multiple developmental processes for more than one outcome variable; sequential-cohort and treatment–control multiple-population studies; and longitudinal analysis for latent variable constructs in the traditional psychometric sense of factor analytic measurement models for multiple indicators. The latent variable framework also accommodates missing data (Arminger & Sobel, 1990; Muthén, Kaplan, & Hollis, 1987), categorical and other nonnormal variable outcomes (Muthén, 1984, 1996), and techniques for clustered (multilevel) data (Muthén, 1994, 1997; Muthén & Satorra, 1995).

SECTION 3: MULTISTAGE GROWTH MODELING OF READING SKILLS DEVELOPMENT USING A CONVENTIONAL LATENT VARIABLE FRAMEWORK

A first attempt at multistage modeling of sequential processes uses the conventional latent variable framework for growth modeling. It is suitable for relating multiple outcome variables to each other. The case of a single outcome variable is discussed first.

Growth modeling with a single outcome variable

Consider a certain outcome variable y_j that is measured repeatedly. For individual i at time t, we may formulate the following linear growth model for this outcome variable:

$$y_{ijt} = \eta_{ij1} + (a_t - a_0)\, \eta_{ij2} + \epsilon_{ijt};\ t = 1, 2, \ldots, T. \tag{4.1}$$

Here, $\eta_{ijk}(k = 1, 2)$ are latent variables, or growth factors, representing the random coefficients of the growth process, the individually varying intercepts and slopes, respectively. Furthermore, a_t denotes a time-related variable such as age, a_0 is an anchor point (such as mean age), and ϵ_{ijt} is a residual. The model may be elaborated by adding time-varying covariates to Equation 4.1 representing educational inputs or other factors influencing the learning at different time points.

The modeling in Equation 4.1 can be used to address the first research hypothesis of Francis (1996): Kindergarten children will differ in their growth and development in precursor skills. The amount of variation in development is captured by the variance of the growth factors η_{ij1} and η_{ij2}. This variation can be explained by background variables observed for the children, such as gender, SES, and ethnicity. A child's developmental status at a given time is of interest when transitioning to a new phase of learning. Here, developmental status refers to the value predicted by the growth curve, not including the time-specific term ϵ_{ijt} in Equation 4.1. For instance, if a_0 represents the end of kindergarten, η_{ij1} represents the developmental status at that time. The child's progress over time adds further useful information. A measure of progress is obtained by η_{ij2}, the linear growth rate for individual i. This describes how the individual reached the kindergarten endpoint. A child may have been close to that level throughout the year or may have experienced rapid growth up to that level. Given an estimated growth model for a sample of individuals, a specific individual's status and growth rate may be estimated by Bayesian methods; in psychometrics, this is termed *factor score estimation*. This describes the essence of how conventional growth modeling can be used to study progress.

Growth modeling with multiple processes

The novel growth modeling feature to be considered is relating the random coefficients of the later process to those of the earlier process. This addresses the second research hypothesis of Francis (1996): The rate of development of the precursor skills will relate to the rate of development and the level of attainment of reading and spelling skills, and individual growth rates in reading and spelling skills will predict performance on standardized tests of reading and spelling.

Phonemic awareness can be taken as an example of a precursor skill. Consider the influence of phonemic awareness on first-grade word recognition. Using the subscripts p and w to replace the generic j subscript in the growth model of Equation 4.1, these outcome variables will be denoted y_{ipt} and y_{iwt} with the corresponding subscripts for the η factors. The intercept and slope equations for the growth coefficients of the first-grade process regressed on those of the kindergarten process may then be written as,

$$\eta_{iw1} = \alpha_1 + \beta_{11}\, \eta_{ip1} + \beta_{12}\, \eta_{ip2} + \zeta_{i1}, \tag{4.2}$$

$$\eta_{iw2} = \alpha_2 + \beta_{21}\, \eta_{ip1} + \beta_{22}\, \eta_{ip2} + \zeta_{i2}. \tag{4.3}$$

Here, the β coefficients represent the strength of the dependencies on past performance and acquired skills in transitioning to a new skill. It is assumed that phonemic awareness development predicts word recognition development, emphasizing the importance of the β transition parameters.

As an additional sequential link, the standardized reading and spelling test scores at the end of first grade can be regressed on the growth coefficients of the first-grade process. Letting the reading and spelling scores be denoted y_r and y_s, respectively,

$$y_r = \alpha_r + \beta_{r1}\, \eta_{iw1} + \beta_{r2}\, \eta_{iw2} + \zeta_{ir} \tag{4.4}$$

$$y_s = \alpha_s + \beta_{s1}\, \eta_{iw1} + \beta_{s2}\, \eta_{iw2} + \zeta_{is}. \tag{4.5}$$

Products of β coefficients in Equations 4.2, 4.3, 4.4, and 4.5 translate progress on precursor skills into predictions of ultimate outcomes on the standardized reading and spelling tests. Background characteristics of the child may have an influence on the dependent variables in all four of these equations.

Assembling the observed variables into the vector $\mathbf{y}_i = (y_{ip1}, \ldots, y_{ipT}, y_{iw1}, \ldots, y_{iwT}, y_{ir}, y_{is})'$ and considering the latent variable vector $\boldsymbol{\eta}_i = (\eta_{ip1}, \eta_{ip2}, \eta_{iw1}, \eta_{iw2}, y_{ir}, y_{is})'$, Equation 4.1 may be fitted into the measurement part of a structural equation model,

$$\mathbf{y}_i = \boldsymbol{\nu} + \boldsymbol{\Lambda}\,\boldsymbol{\eta}_i + \mathbf{K}\,\mathbf{x}_i + \boldsymbol{\epsilon}_i. \tag{4.6}$$

Equations 4.2 through 4.5 may be fitted into the structural part of a structural equation model,

$$\boldsymbol{\eta}_i = \boldsymbol{\alpha} + \mathbf{B}\,\boldsymbol{\eta}_i + \boldsymbol{\Gamma}\,\mathbf{x}_i + \boldsymbol{\zeta}_i, \tag{4.7}$$

where $\mathbf{x}$ represents background variables. The model may be estimated by maximum likelihood under normality assumptions using standard structural equation modeling software (Muthén & Curran, 1997).

Results

The growth model in Equations 4.1, 4.2 and 4.3 was applied to the growth processes of kindergarten phonemic awareness and first-grade word recognition. Linear growth was found to hold for both processes. A sample of $n = 410$ children had complete data on the four kindergarten measures and the four first-grade measures and the analyses are based on these children. To capture the phonemic awareness level at exit from kindergarten, the intercept factor was defined at time point 4. Similarly, the word recognition intercept factor was defined at time point 4 in first-grade.

The maximum-likelihood estimates of the mean of the phonemic awareness slope factor was 0.21. The variance of the intercept and slope factors was 0.64 and 0.02. Both values were significantly different from zero. Their relative size showed the typical feature of a much higher level variation than of a growth rate variation. The correlation between the intercept and slope was high, 0.72. The estimates of the four β coefficients in the growth factor Equations 4.2 and 4.3 are given in Table 4.1.

This indicates that for word recognition level at the end of first grade, represented by the W intercept, the phonemic awareness level at the end of kindergarten (P intercept) is important whereas the kindergarten growth rate (P slope) is insignificant. The amount of variation in the W intercept accounted for by the kindergarten growth factors was 42%. The first-grade

TABLE 4.1
Estimates of the Relations Between the First-Grade and Kindergarten
Growth Factors

Dependent Variable	P Intercept	P Slope
W intercept		
Unstandardized	0.79 (0.07)	-0.41 (0.40)
Standardized	0.70	-0.07
W slope		
Unstandardized	-0.05 (0.02)	0.32 (0.11)
Standardized	-0.24	0.30

Note. Standard errors are in parentheses.

growth rate (W slope) was best predicted by the kindergarten slope (P slope). In this case, however, only 4% of the variation was accounted for.

SECTION 4: MODELING WITH MULTIPLE TRAJECTORY CLASSES

This section describes shortcomings in the analysis of sequential processes using growth modeling in a conventional latent variable framework. An alternative, extended growth model analyzed in a more general latent variable framework is presented.

Shortcomings of the growth model

The growth model allows for individual differences in development. In this way, the estimated model gives not only an estimated mean curve, but also estimates the variation in individual curves as a function of the growth factors. This model allows curves for different individuals to be very different. Nevertheless, the model is restrictive in that it does not recognize that the sample of children may be heterogeneous so that different subgroups may follow different models. This restriction is particularly limiting when attempting to predict a later process from an earlier process.

The use of growth factors as predictors is complicated by the fact that the meaning of a growth factor may be different at different levels of another growth factor. Consider for example the hypothesis that a high kindergarten phonemic awareness intercept and slope interact to influence good first-grade word recognition development. The intercept is defined at the kindergarten exit point so that a high positive slope value means that the child has been at considerably lower levels earlier in kindergarten. This rapid growth can in principle be either good or bad. The rapid growth may be good because the child shows potential for rapid learning that may carry over to first grade. For example, a low starting point in kindergarten may be due to detrimental home circumstances but the child grows because his or her aptitude for reading is good. The rapid growth may be bad because the child has not been at the kindergarten exit level for long and therefore may have had limited learning opportunities during kindergarten. It is conceivable that these two alternatives have different plausibility at different kindergarten exit levels. If this is the case, the influence of the interaction between kindergarten intercept and slope is not monotonic and needs a special modeling approach. An approach of this type is now presented.

Growth mixture modeling

The latent variable model in Equations 4.6 and 4.7 are now modified drawing on the growth mixture model of Muthén, Shedden, and Spisic (2000). This builds on a latent variable structural equation model generalized to K classes of a finite mixture. The heterogeneity of the growth is captured by a categorical latent variable $c_i = (c_{i1}, \ldots, c_{iK})'$, where $c_{ik} = 1$ if individual i falls in class k and zero otherwise. The modeling and estimation is presented first, followed by the application to the reading skills development.

Modeling and Estimation

For each class k, continuous outcome variables $\mathbf{y}$ are assumed normally distributed, conditional on covariates $\mathbf{x}$, related as follows:

$$\mathbf{y}_{ik} = \boldsymbol{\nu}_k + \boldsymbol{\Lambda}_k \, \boldsymbol{\eta}_{ik} + \mathbf{K}_k \, \mathbf{x}_{ik} + \boldsymbol{\epsilon}_{ik}, \tag{4.8}$$

$$\boldsymbol{\eta}_{ik} = \boldsymbol{\alpha}_k + \mathbf{B}_k \, \boldsymbol{\eta}_{ik} + \boldsymbol{\Gamma}_k \, \mathbf{x}_{ik} + \boldsymbol{\zeta}_{ik}. \tag{4.9}$$

The covariance matrices $\boldsymbol{\Theta}_k = V(\boldsymbol{\epsilon}_{ik})$ and $\boldsymbol{\Psi}_k = V(\boldsymbol{\zeta}_{ik})$ are also allowed to vary across the K classes. Here, $\boldsymbol{\alpha}_k$ contains the intercepts for $\boldsymbol{\eta}$ for latent class k. The different $\boldsymbol{\alpha}_k$ values are used to represent different trajectory shapes for the different classes. This is a finite mixture model similar to what was proposed by Verbeke and LeSaffre (1996). To understand membership composition for the different trajectory classes, it is useful to relate the probability of class membership to background variables. As in Muthén and Shedden (1999), a further component is therefore added to the model, where $\mathbf{c}$ is related to $\mathbf{x}$ through a multinomial logistic regression model for unordered polytomous response. Defining $\pi_{ik} = P(c_{ik} = 1|\mathbf{x}_i)$, the K-dimensional vector $\boldsymbol{\pi}_i = (\pi_{i1}, \pi_{i2}, \ldots, \pi_{iK})'$, and the $K - 1$ dimensional vector $logit\ (\boldsymbol{\pi}_i)$ $= (log\ [\pi_{i1}/\pi_{iK}], log\ [\pi_{i2}/\pi_{iK}], \ldots, log\ [\pi_{i,K-1}/\pi_{iK}])'$, this model part is expressed as

$$logit\ (\boldsymbol{\pi}_i) = \boldsymbol{\alpha}_c + \boldsymbol{\Gamma}_c\ \mathbf{x}_i, \tag{4.10}$$

where $\boldsymbol{\alpha}_c$ is a $K - 1$ dimensional parameter vector and $\boldsymbol{\Gamma}_c$ is a $(K - 1) \times q$ parameter matrix.

Maximum-likelihood estimation under normality assumptions can be carried out using the EM algorithm. In the EM algorithm, data are considered missing on the latent categorical variable $\mathbf{c}_i$. The complete-data likelihood of the EM algorithm for the model in Equations 4.8, 4.9, and 4.10 considers

$$[\mathbf{c}|\mathbf{x}]\ [\mathbf{y}|\mathbf{c}, \mathbf{x}], \tag{4.11}$$

where $[\mathbf{z}]$ denotes a density or probability distribution. The first term of Equation 4.11 corresponds to a multinomial regression with a multinomial latent categorical dependent variable determined by Equation 4.10, whereas the second term corresponds to a multivariate normal distribution, $f(\mathbf{y}_{ik}|\mathbf{x}_i) = N(\boldsymbol{\mu}_{ik}, \boldsymbol{\Sigma}_k)$ derived from Equations 4.8 and 4.9. The E and M steps of the algorithm are discussed in Muthén, Shedden, and Spisic (2000). A useful side product of the analysis is estimates of posterior probabilities for each individual's class membership,

$$p_{ik} = P(c_{ik} = 1|\mathbf{y}_i, \mathbf{x}_i) \propto P(c_{ik} = 1|\mathbf{x}_i)\ f(\mathbf{y}_{ik}|\mathbf{x}_i). \tag{4.12}$$

An individual may be classified into the class for which he or she has the highest posterior probability.

In the context of growth modeling, the finite mixture model is referred to as a growth mixture model. Mixture modeling can be viewed as a form of cluster analysis. Many researchers have attempted to cluster longitudinal measures to capture different classes of trajectories by various ad hoc methods. The present method is a rigorous parametric approach; for related mixture approaches to clustering, see, McLachlan and Basford (1988). In the present study, a "confirmatory" clustering approach is used, where parameter restrictions are imposed based on a priori hypotheses about growth. Different prespecified growth shapes can be captured by letting some of the parameters of $\boldsymbol{\alpha}_k$ be fixed. The growth mixture modeling results shown later were obtained using the new latent variable modeling software Mplus (Muthén & Muthén, 1998). Input specifications for the analyses can be obtained from the first author.

The posterior probability computations shown in Equation 4.12 can be used to derive the most likely class membership for a given individual observation vector $(\mathbf{y}_i, \mathbf{x}_i)$. A typical use is where the estimated model is taken as given and a new individual from the same population is observed. Here, the estimated model is used as a measurement instrument in the sense that an observation vector is translated into a class membership statement. The Mplus program can be used for such posterior probability calculations holding all model parameters fixed at the estimated values and doing only one E step. Because the estimated model is still valid for a subset of the outcome variables in $\mathbf{y}_i$, posterior probabilities can also be computed using a subset of the repeated measures on $\mathbf{y}_i$ up to a certain time point. This responds to questions of how early a useful classification can be obtained.

The growth mixture modeling approach also provides a way to study early indications of problematic development. As an example, it is of interest to be able to identify students who are likely to belong to Class 1. The estimated posterior probabilities obtained by Equation 4.12 provides a classification of each individual into the class with the highest probability. This is of interest when using the estimated model to classify a new student as early as possible. In this case, the parameters of the estimated model are taken as given and only the posterior probabilities are estimated. Although the model is estimated from all the y and x variables, the estimation of the posterior probabilities can be done using only a subset of early measurements. This is a useful approach to identify children who are at risk for reading failure as early as possible. Muthén, Francis, and Boscardin (1999) provides an analysis of this kind.

Application to reading skills development

Applied to the prediction of first-grade word recognition growth using kindergarten phonemic awareness growth, $\mathbf{y}_i = (y_{ip1}, \ldots, y_{ip4}, y_{iw1}, \ldots,$ $y_{iw4}, y_r, y_s)'$ and $\boldsymbol{\eta}_i = (\eta_{ip1}, \eta_{ip2}, \eta_{iw1}, \eta_{iw2}, \eta_r, \eta_s)'$. Here, the modeling includes the standardized reading and spelling test scores y_r and y_s at the end of first grade. These scores are included in the model as two further η variables η_r and η_s that are perfectly measured by corresponding y variables ($\epsilon_r = 0, \epsilon_s = 0$). To illustrate the use of covariates $\mathbf{x}$ in Equation 4.10, a measure of letters, name, and sounds skills obtained at the beginning of kindergarten is used. This serves as a proxy for home literacy support and early instruction and is a rudimentary early indicator of both automation of the symbol recognition process needed for deciphering print into language and, in the case of letter sounds, of phonemic awareness/grapho–phonemic awareness.

In Equation 4.9, the first two elements of $\boldsymbol{\alpha}_k$ contain the means of the phonemic awareness intercept and slope and the next two elements contain the means of the word recognition intercept and slope. The trajectory classes are obtained by fixing the $\boldsymbol{\alpha}_k$ mean of the kindergarten phonemic awareness intercept and slope to different values. Four classes are chosen to represent variation in both intercept and slope values for phonemic awareness development; they are described later. The latent class variable is a predictor of first-grade development of word recognition. This is expressed by Equation 4.9 where the six vectors α_k capture the across-class differences in means. The estimated values of the word recognition intercept and slope means in α_k are of primary interest in the analysis. Given the high number of classes, it is assumed that relatively little within-class variation remains in these growth factors. The variation is instead represented by the latent classes. For simplicity, the latent class variable is therefore taken as the only predictor of first-grade development of word recognition with corresponding zero elements of **B** in Equation 4.9 in the present analysis. The η_r and η_s variables are specified to be predicted by the latent class variable in the sense that their means are allowed to vary across classes, and they will also be predicted by the intercept factors for phonemic awareness and word recognition with corresponding nonzero elements in **B**. The model is shown in path diagram form in the bottom part of Fig. 4.1, where, as a comparison, the top part represents the conventional growth model estimated in Section 3.

The four prespecified trajectory classes for phonemic awareness are shown in the left-hand panel of Fig. 4.2. Each line is plotted at the mean values of the phonemic awareness intercept and slope for the class. Each class allows variation around this line as a function of variation in the intercept and slope. The classes represent three different mean values at the exit of kindergarten. These values are determined from the mean and variance of the growth intercept in a single-class analysis of these data, where the intercept is defined at the end of kindergarten. The values are the mean and plus and minus one standard deviation away from the mean of the intercept growth factor. The slopes for all classes except Class 1 are the average values given that intercept value. Classes 1 and 2 differ only in the growth slope, where Class 1 has zero growth. Class 1 is of special interest given that it shows failure in reading precursor development. It is also of interest to contrast Class 1 with Class 2. The choice of four classes is not based on model fit criteria but on the degree of separation of classes that is of substantive interest and that can be supported by the analysis. In earlier analyses, six classes were used but two classes gave zero class counts when analyzing the full model.

It is of interest to be able to identify students who are likely to belong to the different classes. The estimated posterior probabilities obtained by Equation 4.12 provides a classification of each individual into the class with the highest probability. In this case, the parameters of the estimated model are taken as given and only the posterior probabilities are estimated.

FIG. 4.1. Path diagrams for conventional growth modeling and growth mixture modeling.

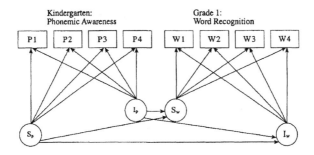

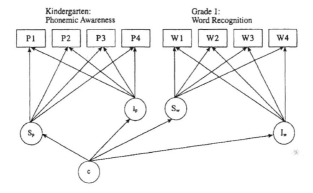

Growth mixture results

Growth mixture analysis was applied to the same sample as in the previous example, except it was reduced to $n = 409$ due to the inclusion of the covariate. In the analysis, initial specifications of class-invariant parameters were relaxed stepwise to see if a solution could be found with a significantly better fit. Here, fit was evaluated by a log likelihood ratio chi-square statistic obtained for nested models. Growth factor variances for the kindergarten phonemic awareness intercept and slope were found to be class varying with a particularly large intercept variance for Class 4. The residual variances for the standardized reading and spelling test scores were also found to be class varying.

Table 4.2 shows the prespecified means for the phonemic awareness

FIG. 4.2. Estimated mean trajectories for growth mixture model.

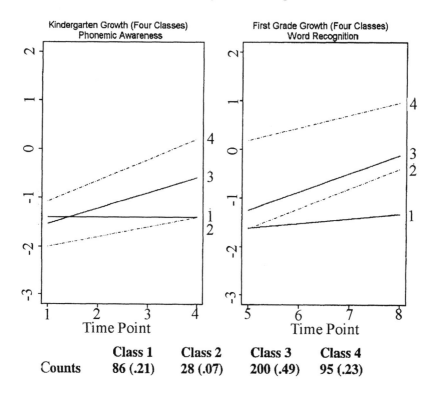

	Class 1	Class 2	Class 3	Class 4
Counts	86 (.21)	28 (.07)	200 (.49)	95 (.23)

intercept and slope for the four classes and also the estimated class probabilities. It is seen that Class 1, showing no kindergarten growth and a low level at exit from kindergarten, contains 21% of the children. Class 2, showing rapid kindergarten growth but the same low level at exit from kindergarten contains 7% of the children. Class 3 and Class 4 contain children with average level and above average level at exit from kindergarten, respectively, with 49% and 23%.

Table 4.3 shows the estimated word recognition intercept and slope means for the four classes. The corresponding estimated trajectories are shown in the right-hand part of Fig. 4.2.

Table 4.3 and Fig. 4.2 show that children in Class 1 continue to do poorly during first grade in terms of word recognition development. Children in Class 1 do better than children in Class 2. This responds to the earlier discussion about whether rapid growth up to a certain level is

TABLE 4.2

Fixed Values for the kindergarten Phonemic Awareness Intercept and
Slope Means and the Estimated Class Probabilities.

	Intercept	Slope	Probability
Class 1	-1.40	0.00	.21
Class 2	-1.40	0.20	.07
Class 3	-0.59	0.32	.49
Class 4	0.20	0.43	.23

TABLE 4.3

Estimated Values for the First-Grade Word Recognition Intercept and
Slope Means

	Intercept	Slope
Class 1	-1.33 (.13)	0.10 (.05)
Class 2	-0.41 (.20)	0.41 (.04)
Class 3	-0.12 (.06)	0.38 (.02)
Class 4	0.95 (.06)	0.26 (.02)

Note. Standard errors are in parentheses.

better than having been at that level longer. These results indicate that at
this kindergarten exit level, rapid growth is preferable for good first-grade
development of word recognition. Children in Class 3 and Class 4 continue
to do well in first grade.

The standardized reading test score was found significantly related
to the word recognition intercept and slope, but not to the phonemic
awareness intercept or slope. The standardized spelling test score was
found significantly related to the phonemic awareness intercept and the
word recognition intercept. As expected, the estimated means of the two
test scores were in increasing order for Class 1, Class 2, Class 3, and Class
4. For both of the two test scores, the Class 1 mean was estimated at
approximately one standard deviation below the overall mean.

The estimated multinomial regression of the latent class variable on the
letters, sounds, and names covariate is summarized in Fig. 4.3. The mean

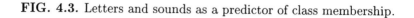

FIG. 4.3. Letters and sounds as a predictor of class membership.

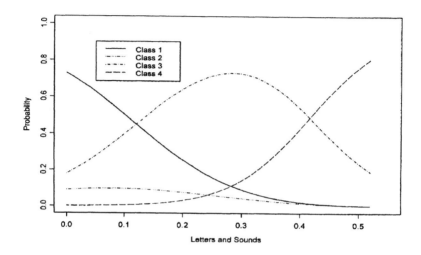

and variance of this covariate are 0.27 and 0.16, respectively. The figure shows that for students who have covariate values lower than one standard deviation below the mean, Class 1 membership is most likely. For increasing covariate values Class 3 and Class 4, respectively, become more likely.

CONCLUSIONS

The general growth mixture modeling approach was found to be a useful tool for studying the relationship between two sequential processes. It avoided the complexity of predicting growth in the later process by the growth factors of the earlier process. Instead, a latent class variable with classes corresponding to prespecified growth shapes was used to predict growth in the later process.

Application to predicting first-grade word recognition development by kindergarten phonemic awareness development resulted in several interesting findings. In particular, we found that among children with low phonemic awareness scores at the end of kindergarten, those who had shown little growth during kindergarten continued to do poorly in terms of word recognition during first grade. An estimated 21% of the children in this sample showed this type of development. The children who had started out lower but had grown rapidly up to this low phonemic awareness level at the end of kindergarten performed significantly better in terms of word recognition during first grade. An estimated 7% showed this type of development.

The results from the growth mixture analysis may be contrasted with those of the conventional, single-class analysis in Section 3.   In the single-class analysis, the slope of the phonemic awareness development was not found to be a significant predictor of the word recognition intercept at exit from first-grade. In contrast, the growth mixture analysis showed that the phonemic awareness slope is an important determinant of word recognition level at exit from first-grade as illustrated by comparing word recognition development for Class 1 and Class 2.

The growth mixture modeling approach also provides a way to study early indications of problematic development. For example, it is of interest to be able to identify students who are likely to belong to Class 1. The estimated posterior probabilities provide a classification of each individual into the class with the highest probability. This is of interest when using the estimated model to classify a new student as early as possible. In this case, the parameters of the estimated model are taken as given and only the posterior probabilities are estimated. Although the model is estimated from all the observed variables, the estimation of the posterior probabilities can be done using only a subset of early measurements. This is a useful approach to identify children who are at risk for reading failure as early as possible. Muthén et al. (1999) provides an analysis of this kind.

The line of research described here has important implications for preventive interventions and choice of treatment; in other words, different methods of teaching reading. Children belonging to different trajectory classes may respond differently to a given treatment and the modeling can be used to better assess treatment–aptitude interactions. The modeling can also be used to design different treatments for children belonging to different trajectory classes.

ACKNOWLEDGMENTS

The work reported herein was supported under the Educational Research and Development Center Program, PR/Award Number R305B60002, as administered by the Office of Educational Research and Improvement, U. S. Department of Education, and by grant HD28172 from the National Institute of Child Health and Human Development. The findings and opinions expressed in this report do not reflect the positions or policies of the National Institute on Student Achievement, Curriculum, and

Assessment, the Office of Educational Research and Improvement, or the U. S. Department of Education. We thank Jack M. Fletcher and Barbara R. Foorman for access to their data and Noah Yang for helpful research assistance.

References

Arminger, G., & Sobel, M. E. (1990). Pseudo-maximum likelihood estimation of mean and covariance structures with missing data. *Journal of the American Statistical Association, 85,* 195–203.

Bryk, A. S., & Raudenbush, S. W. (1992). *Hierarchical linear models: Applications and data analysis methods.* Newbury Park, CA: Sage.

Francis, D. (1996). Detecting reading problems by modeling individual growth (Grant proposal).

Goldstein, H. (1995). *Multilevel statistical models.* London: Edward Arnold.

Jennrich, R. I., & Schluchter, M. D. (1986). Unbalanced repeated-measures models with structured covariance matrices. *Biometrics, 42,* 805–820.

Laird, N. M., & Ware, J. H. (1982). Random-effects models for longitudinal data. *Biometrics, 38,* 963–974.

Lindstrom, M. J., & Bates, D. M. (1988). Newton-Raphson and EM algorithms for linear mixed-effects models for repeated-measures data. *Journal of the American Statistical Association, 83,* 1014–1022.

MacCallum, R. C., Kim, C., Malarkey, W. B., & Kiecolt-Glaser, J. K. (1997). Studying multivariate change using multilevel models and latent curve models. *Multivariate Behavioral Research.*

McLachlan, G. J., & Basford, K. E. (1988). *Mixture models: Inference and applications to clustering.* New York: Marcel Dekker.

Meredith, W., & Tisak, J. (1990). Latent curve analysis. *Psychometrika, 55,* 107–122.

Muthén, B. O. (1984). A general structural equation model with dichotomous, ordered categorical, and continuous latent variable indicators. *Psychometrika, 49,* 115–132.

Muthén, B. O. (1994). Multilevel covariance structure analysis. *Sociological Methods and Research: Multilevel Modeling* [Special issue], In J. Hox & I. Kreft (Eds.), *22,* 376–398.

Muthén, B. O. (1996). Growth modeling with binary responses. In A. V. Eye & C. Clogg (Eds.), *Categorical variables in developmental research: Methods of analysis* (pp. 37–54). San Diego, CA: Academic Press.

Muthén, B. O. (1997). Longitudinal and multilevel modeling. In A. E. Raftery (Ed.), Sociological Methodology (pp. 453–480). Boston: Blackwell.

Muthén, B. O., & Curran, P. J. (1997). General longitudinal modeling of individual differences in experimental designs: A latent variable framework for analysis and power estimation. *Psychological Methods, 2,* 371–402.

Muthén, B., Francis, D., & Boscardin, C. K. (1999). Early prediction of reading skills trajectory class. Manuscript in preparation.

Muthén, B. O., Kaplan, D., & Hollis, M. (1987). On structural equation modeling with data that are not missing completely at random. *Psychometrika, 42,* 431–462.

Muthén, B. O. & Muthén, L. K. (1998). Mplus user's guide. Los Angeles, CA: Authors.

Muthén, B. O., & Satorra, A. (1995). Complex sample data in structural equation modeling. In P. V. Marsden (Ed.), *Sociological methodology* (pp. 267–316). Washington, DC: American Sociological Association.

Muthén, B., & Shedden, K. (1999). Finite mixture modeling with mixture outcomes using the EM algorithm. *Biometrics, 55,* 463–469.

Muthén, B., Shedden, K., & Spisic, D. (2000). *General latent variable mixture modeling.* Technical report.

Thum, Y. M. (1997). Hierarchical linear models for multivariate outcomes. *Journal of Educational and Behavioral Statistics, 22*(1), 76–107.

Verbeke, G., & LeSaffre, E. (1996). A linear mixed-effects model with heterogeneity in the random effects population. *Journal of the American Statistical Association, 91,* 217–221.

5

Multilevel Models for Meta-Analysis

Joop J. Hox
Utrecht University, Utrecht

Edith D. de Leeuw
MethodikA, Amsterdam

Meta-analysis concerns the statistical integration of a large number of results from empirical studies (cf. Glass, 1976). The goal is to summarize the results of a collection of independently conducted studies on one specific research question. For instance, the research question might be: What is the effect of social skills training on socially anxious children? In a meta-analysis, one would collect reports of experiments concerning this question, explicitly code the reported outcomes, and integrate the outcomes statistically into a combined "super outcome". Often the focus is not on integrating or summarizing the outcomes, but on more detailed questions about variations in the outcomes, such as: What is the effect of different durations for the training sessions? Are there differences between different training methods? In these cases, the meta-analyst not only examines the overall study outcomes, but also codes study characteristics. These study characteristics, for example design features or type of subjects sampled, are potential explanatory variables to explain differences in the study outcomes.

The core of meta-analysis is that statistical analyses are carried out on the published results of a collection of empirical studies on a specific research question. A very general model for meta-analysis is the random effects model (Hedges & Olkin, 1985). In this model, the focus is on analyzing the size of the effects found in the different studies, not on establishing the statistical significance of a combined outcome. The random effects model assumes that study outcomes vary not only because of random sampling effects (variations within each study), but also because of real differences between the studies. For instance, study outcomes can vary because different studies employ different sampling methods, use different experimental manipulations, or measure the effects with different instruments. The random effects model is used to decompose the variance of the study outcomes into a part that is the result of sampling variation, and a part that reflects real differences between the studies. Hedges and Olkin (1985) gave procedures that can be used to decompose the total variance of the study outcomes into random sampling variance and systematic between-study variance, and to test the significance of the between-study variance. If the between-study variance is large and significant, the study outcomes are *heterogeneous*. In that case, the usual procedure is to form clusters of studies that differ in their outcomes, but that are homogeneous within the clusters. These clusters can be constructed a priori, based on available study characteristics; they can also be constructed a posteriori, based on a cluster analysis of the reported outcomes. The goal is to identify study characteristics that explain differences between the study outcomes. Variables that affect the study outcomes are in fact moderator variables, that is, variables that interact with the independent variable.

Meta-analysis can be viewed as a special case of multilevel analysis. We have a hierarchical data set, with subjects within studies at the first level, and studies at the second level. If the raw data of all the studies was available, we could carry out a standard multilevel analysis, predicting the outcome variable using the available individual and study–level explanatory variables. In our example, we would have one outcome variable, for instance the result on a test measuring social skill, and one explanatory variable, which is a dummy variable that indicates whether the subject is in the experimental or the control group. On the individual level, we have a linear regression model that relates the outcome to the grouping variable. The general multilevel regression model assumes that each study has its own regression model. In particular, the intervention effect (e.g., the regression coefficient for the grouping variable in each study) is allowed to vary across studies. Standard multilevel analysis can be used to estimate the mean and variance of the intervention effects across the studies. If the intervention effects vary substantially and significantly across studies, we have heterogeneous results. In that case, we can study further the variation of intervention effects across studies by examining study-level regression models that attempt to explain the study-specific intervention effects with the available study characteristics as explanatory variables.

These analyses can be carried out using standard multilevel regression methods (cf. Bryk & Raudenbush, 1992; Goldstein, 1995; Hox, 1995; Snijders & Bosker, 1999) and standard multilevel software. A special complication is that in meta-analysis we usually do not have access to the original raw data. Instead, we have the published results in the form of p-values, means, standard deviations, or correlation coefficients. Classical meta-analysis has developed a large variety of methods to integrate these statistics into one overall outcome. Hunter and Schmidt (1990) discussed these methods in detail, and Hedges and Olkin (1985) discussed the statistical models used.

Nevertheless, it is possible to carry out a multilevel meta-analysis on the data that are usually available in meta-analysis. Raudenbush and Bryk (Bryk & Raudenbush, 1992; Raudenbush & Bryk, 1985) presented the random effects model for meta-analysis as a special case of the multilevel regression model. The analysis is performed on sufficient statistics instead of raw data, and as a result, some specific restrictions must be imposed on the model. Analytic procedures for the standard multilevel software HLM and MlwiN are given in the Appendix. The major advantage of using multilevel analysis instead of classical meta-analysis methods is flexibility. In multilevel meta-analysis, it is simple to include study characteristics as explanatory variables in the model. If we have hypotheses about study characteristics that influence the outcomes, we can code these and include them on a priori grounds in the analysis. Alternatively, after we have concluded that the study outcomes are heterogeneous, we can explore the available study variables in an attempt to explain the heterogeneity.

THE MODEL

In a typical meta-analysis, the studies usually employ different instruments and use different statistical tests. To make the outcomes comparable, the study results must be transformed into a standardized measure of the effect, such as a correlation coefficient or the standardized difference between two means, d. The general model for the study outcomes is

$$d_j = \delta_j + e_j. \tag{5.1}$$

In Equation 5.1, d_j is the outcome of study j $(j=1,\ldots,J)$, δ_j is the population value of this outcome, and e_j is the sampling error for this study. It is assumed that the e_j have a normal distribution with a known variance σ_j^2. If the sample sizes of the individual studies are not too small, for instance 20 (Hedges & Olkin, 1985) to 30 (Bryk & Raudenbush, 1992), the assumption that the sampling distribution of the outcomes is normal is usually reasonable. Most classical meta-analysis methods also assume normality (cf., Hedges & Olkin, 1985). The variance of the sampling distribution is often known from statistical theory; in some cases, a transformation is needed to achieve normality and known sampling

TABLE 5.1

Effect Measures and Their Sampling Variance

Measure	Estimator	Transformation	Sampling variance
Mean	$\bar{X}$	-	$\frac{s^2}{n}$
Diff. 2 Means	$g = \frac{(\bar{Y}_E - \bar{Y}_C)}{s}$	-	$\left(\frac{n_E + n_C}{n_E n_C}\right) + \left(\frac{g^2}{2n_E + n_C}\right)$
St. Dev.	s	$s* = LN(s) + \frac{1}{2}df$	$\frac{1}{2}df$
Correlation r		$z = .5LN\left[\frac{(1+r)}{(1-r)}\right]$	$1/(n-3)$
Proportion p		$l* = LN[p/(1-p)]$	$1/[np(1-p)]$

variance. Table 5.1 lists some common effect size measures; if needed, the normalizing transformation; and the corresponding sampling variance.

When using the sample mean as the effect measure, we should make sure that the outcomes are comparable across studies. If different outcome measures are used, the measures might be scaled in different units. Without some kind of standardization, comparing those outcomes is like comparing apples and oranges, or rather, like comparing pounds and kilograms.

In Table 5.1, g is the effect size proposed by Glass (1976). The transformation of s to $s*$ for the standard deviation is proposed by Bryk and Raudenbush (1992). The transformation for the correlation r is the familiar Fisher-Z transformation, and for the proportion, the logit. Usually, if a confidence interval is constructed for the transformed variable, the endpoints are translated back to the original estimator. For a more extended list of effect size measures and their sampling variance, see Rosenthal (1994) and Cornell and Mulrow (1999).

Equation 5.1 shows that the parameters δ_j, the study outcomes, are assumed to vary across the studies. The variation of δ_j is assumed to follow the regression model

$$\delta_j = \gamma_0 + \gamma_1 Z_{1j} + \gamma_2 Z_{2j} + \ldots + \gamma_p Z_{pj} + u_j, \qquad (5.2)$$

where $Z_1 \ldots Z_p$ are study characteristics, $\gamma_1 \ldots \gamma_p$ are the regression coefficients, and u_j is the residual error term, which is assumed to have a normal distribution with variance σ_u^2. In meta-analysis, there are typically two kinds of study characteristics that can be used as explanatory variables in the regression model; methodological characteristics like study size, methodological quality, and reliability of instruments, and variables that are of theoretical interest, such as the type and intensity of intervention, or duration of the intervention.

By substituting Equation 5.2 into Equation 5.1 we get the complete model

$$d_j = \gamma_0 + \gamma_1 Z_{1j} + \gamma_2 Z_{2j} + \ldots + \gamma_p Z_{pj} + u_j + e_j. \qquad (5.3)$$

If we have no explanatory variables, the model reduces to

$$d_j = \gamma_0 + u_j + e_j. \qquad (5.4)$$

Equation 5.4, which in multilevel analysis is often denoted as the "intercept only" or "null" model (cf., Bryk & Raudenbush, 1992), is equivalent to the random effects model for meta-analysis described by Hedges and Olkin (1985). Hedges and Olkin described a one-step weighted least squares procedure for estimating the model parameters. Multilevel analysis programs typically use iterative maximum likelihood estimation, which, in general, is more efficient (Raudenbush, 1994). In practice, both models usually produce very similar parameter estimates.

In Equation 5.4, the intercept γ_0 is the estimate for the mean outcome $\bar{\delta}$ across all studies. The variance of the outcomes (δ_j) across studies, σ_u^2, indicates how much the studies' outcomes vary. Thus, to test if the study outcomes are homogeneous is equivalent to testing the null hypothesis that σ_u^2 is equal to zero. If the test of σ_u^2 is significant, the study outcomes are heterogeneous.

Note that Equation 5.4 contains two residual error terms; u_j and e_j. The variance of the u_j, σ_u^2, represents the true variation between the studies, which is estimated in the meta-analysis, and which we would like to explain using the available study characteristics. The e_j represents the differences between the studies that are the result of sampling variation. The sampling variance of the studies, $\sigma_{e_j}^2$, is determined fully by the within-study variation and sample size, and assumed known from the study's publications. Consequently, the sampling variance, calculated from the published results, is part of the data to be input in the program (software implementations for multilevel meta-analysis are discussed in an appendix to this chapter). Because the $\sigma_{e_j}^2$ are directly input as known data, there is no assumption that they are homogeneous, that is, that all $\sigma_{e_j}^2$, are equal. Typically, σ_e^2, which is the (weighted) average sampling variance, is estimated by subtracting σ_u^2 from the total variance between studies. The proportion of between-study variance is estimated by the intraclass correlation, which can be estimated using the intercept-only (null) model as $\rho = \sigma_u^2 / (\sigma_u^2 + \sigma_e^2)$. The proportion of systematic between-study variance can be used as an additional indicator of the degree of heterogeneity of the study outcomes. This is analogous to using the proportion of explained variance in standard regression models to indicate the importance of specific predictor variables. Hunter and Schmidt (1990) pointed out that with a large number of studies, the power of the significance test is high, and small variances will become significant. When the number of studies is small, lack of significance for σ_u^2 does not imply that the outcomes are homogeneous. Hunter and Schmidt (1990) proposed a 25% rule of thumb; that is, if the between-study variance is 25% or more of the total variance, it is interesting enough to merit exploration. In our terminology, if the intraclass correlation ρ is 0.25 or higher, the variance between studies is deemed large enough to attempt to model it using the available study characteristics.

The general Equation 5.3 includes study characteristics Z_{pj} to explain differences in the studies' outcomes. In Equation 5.3, σ_u^2 is the residual between-study variance after the explanatory variables are included in the model. The statistical test on σ_u^2 now tests whether the explanatory variables in the model explain all the variation in the studies' outcomes, or if there is still unexplained between-study variance left in the outcomes. The difference between the between-studies variance σ_u^2 in the null model and in the model that includes the explanatory variables Z_{pj}, can be interpreted as the amount of variance explained by the explanatory variables, that is, by the study characteristics included in Equation 5.3.

The multilevel meta-analysis model given by Equation 5.3 is similar to the general model for fixed effects as described by Hedges and Olkin (1985, chap. 8). In our notation, their model is given by

$$d_j = \gamma_0 + \gamma_1 Z_{1j} + \gamma_2 Z_{2j} + \ldots + \gamma_p Z_{pj} + e_j. \tag{5.5}$$

Compared to Equation 5.3, Equation 5.5 lacks the study-level residual error term u_j. Thus, the general model for fixed effects described by Hedges and Olkin is a special case of the multilevel meta-analysis model. Omitting the study-level residual error term u_j implies the assumption that the explanatory variables in the model explain all of the variance across the studies. There are situations, for instance when we limit our statistical generalization to the set of studies at hand, where the fixed effects model is appropriate (Hedges & Vevea, 1998). If this is the case, we can model differences between studies using fixed effects (Overton, 1998). Alternatively, we may find empirically that the studies are homogeneous, meaning that the estimate for the between-study variance σ_u^2 is small and insignificant.

However, in general, we do want to generalize beyond the specific set of studies at hand, and experience shows that usually we cannot explain all between-study variance, just as in ordinary multiple regression analysis, we seldom find a multiple correlation coefficient equal to one. Hunter and Schmidt (1990) assumed that between-studies heterogeneity is partly due to a large number of possible artifacts in the meta-analysis. An example of such an artifact is the (usually untestable) assumption of a normal distribution for the sampling errors e_j. However, unless the sample size is very small in some studies, the normality assumption for e_j is usually reasonable by central limit theorem. Other artifacts that may cause variation between the studies are the correctness of the statistical assumptions made in the original analyses, differences in the reliability of the instruments used in different studies, coder unreliability in coding study characteristics, and so forth. It is unlikely that the available study–level variables cover all of these artifacts. Generally, the amount of detail in the input for the meta-analysis, which are the research reports, papers, and articles, is not enough to cover all of these study characteristics. Therefore, to some extent, heterogeneous results are to be expected. It is this reasoning that led Hunter and Schmidt to their rule of thumb that the between-study variance in the null model should be larger than 25% of the totel variance. The same reasoning led us to the conclusion that, in general, random effects models, such as multilevel regression models, should be used in meta-analysis.

Overton (1998) examined the differences between fixed and random effects models for meta-analysis using simulation. He found, not surprisingly, that fixed effects models perform best when data are generated following a fixed model, and random effects models perform best when data are generated following a random model. Because the fixed effects model is a special case of the random effects model, the best analysis strategy appears to be to begin by estimating a random effects model. If the between-study variance σ_u^2 turns out to be insignificant and negligible in size, the between-study variance can be fixed at zero, which effectively turns the multilevel analysis into a fixed effects analysis.

EXAMPLE AND COMPARISON WITH CLASSICAL META-ANALYSIS

In this section, we analyze an example data set using classical meta-analysis methods as implemented in the program META by Schwarzer (1989). This program is based on methods and procedures described by Rosenthal (1984), Hunter and Schmidt (1990), and Hedges and Olkin (1985). The (constructed) data set consists of 20 studies that compare an experimental group and a control group.[1]

---

[1] The example data were constructed using a regression model like Equation 5.4 with a single explanatory variable "duration in weeks", which was simulated from a normal distribution ($\mu = 6$, $\sigma = 2.5$). The population effect size δ for each study was predicted using a regression slope of 0.15 for the duration with mean outcome 0.6 across all studies.

If we compare the means of an experimental and a control group, an appropriate outcome measure is the standardized difference between the experimental and the control group, originally proposed by Glass (1976) and defined by Hedges and Olkin as $g = (\bar{Y}_E - \bar{Y}_C)/s$, where s is the pooled standard deviation of the two groups. Because g is not an unbiased estimator of the population effect $\delta = (\mu_E-\mu_C)/\sigma$, Hedges and Olkin preferred a corrected effect measure d: $d = \{1-3/[4(n_E+n_C)-9]\}g$. The sampling variance of the effect estimator d is equal to $(n_E + n_C)/(n_E n_C) + d^2/[2(n_E + n_C)]$ (Hedges & Olkin, 1985, p. 86).

Table 5.2 lists both g and d for all 20 studies. With commonly used sample sizes, the difference between the two is very small. Table 5.2 also presents the sampling variance of $d[\text{var}(d)]$, the one-sided p-value of the t test for the difference between the two means (p), the number of cases in the experimental (n_{exp}) and control group(n_{con}), and the reliability (r_{ii}) of the outcome measure used in the study. The example data set contains one study–level explanatory variable, the duration in number of weeks of the experimental intervention. It is plausible to assume that longer interventions lead to a larger effect. In Table 5.2, the studies are presented in increasing order of their effect sizes (g, d).

CLASSICAL META-ANALYSIS

Classical meta-analysis contains a variety of approaches that complement each other. An old approach is to combine the p values of the studies into one overall p value for the collection of studies. Several formulas are available for combining p-values. A popular procedure is the so-called Stouffer method (see Rosenthal, 1984). Each individual p is converted to the corresponding standard normal Z score. The Z scores are then combined using $Z = (\sum Z_i)/\sqrt{k}$, where Z_j is the Z score of study j, and k is the number of studies. For our example, the Stouffer method gives a combined Z of 7.73, which is highly significant ($p < 0.001$).

The combined p value gives us proof that an effect exists, but no information on the size of the experimental effect. The next step in classical meta-analysis is to combine the effect sizes of the studies into one overall effect size, and to establish the significance or a confidence interval for the combined effect. Considering the possibility that the effects may differ across the studies, the random effects model is used to combine the studies.

The sample sizes for the experimental and control group were generated independently from a normal distribution ($\mu = 30$, $\sigma = 10$), and the reliability was randomly chosen from the values 0.9 and 0.75. The reliability was used to attenuate the effect size δ, and finally, the observed effect size g was simulated by adding to δ, a random residual drawn from a normal distribution with mean 0 and variance determined by the sample sizes of the experimental and control groups.

TABLE 5.2

Example Results From Twenty Studies

Study	Duration	g	d	$var(d)$	p	n_{exp}	n_{con}	r_{ii}
1	3	.268	.264	.086	.810	23	24	.90
2	1	.235	.230	.106	.756	18	20	.75
3	2	.168	.166	.055	.243	33	41	.75
4	4	.176	.173	.084	.279	26	22	.90
5	3	.228	.225	.071	.204	29	28	.75
6	6	.295	.291	.078	.155	30	23	.75
7	7	.312	.309	.051	.093	37	43	.90
8	9	.442	.435	.093	.085	35	16	.90
9	3	.488	.476	.149	.116	22	10	.75
10	6	.628	.617	.095	.030	18	28	.75
11	6	.660	.651	.110	.032	44	12	.75
12	7	.725	.718	.054	.003	41	38	.90
13	9	.751	.740	.081	.009	22	33	.75
14	5	.756	.745	.084	.009	25	26	.90
15	6	.768	.758	.087	.010	42	17	.90
16	5	.938	.922	.103	.005	17	29	.90
17	5	.955	.938	.113	.006	14	31	.75
18	7	.976	.962	.083	.002	28	26	.90
19	9	1.541	1.522	.100	.0001	50	16	.90
20	9	1.877	1.844	.141	.00005	31	14	.75

A meta-analysis of the effect sizes in Table 5.2, using the random effects model, estimates the overall effect as $\delta = 0.58$, with a standard error of 0.11. This gives us a Z value of 5.27 ($p < 0.001$). The 95% confidence interval for the overall effect size is $0.36 < \delta < 0.80$. The usual significance test of the variance is a chi-square test on the residuals, which for our example data leads to $\chi^2(19) = 48.9$, $p < 0.001$. As this is clearly significant, we have heterogeneous outcomes. This means that the overall effect 0.58 is not the estimate of a fixed population value, but a (weighted) *average* of the distribution of effects in the population.

The between-study variance σ_u^2 is estimated as 0.17 and the proportion of between-study variance as 0.65. This is much larger than the 0.25 threshold that Hunter and Schmidt (1990) recommended for examining differences between studies. The usual approach in classical meta-analysis is to divide the studies into clusters that have different average effect sizes, while being internally homogeneous. A unidimensional cluster analysis of the study outcomes can be used to form such groups. In our example, a cluster analysis produces three clusters. The first cluster consists of Studies 1 and 2, the second cluster consists of Studies 3 through 18, and the third cluster consists of Studies 19 and 20. For a post hoc interpretation of the clusters, we must examine how the clusters differ. If we look at the mean duration of the experiment in the three clusters, we find that this is 2 weeks in the first cluster, 6 weeks in the second, and 9 weeks in the third. Apparently, the duration of the experimental intervention indeed affects the study outcome.

Because we have the hypothesis that the duration of the experimental intervention is related to the outcome, we can also form a priori clusters based on this variable. We distinguish two a priori clusters; the first consists of the 9 studies that have a duration of 5 weeks or less, and the second consists of the 11 studies that have a duration of 6 weeks or more. The overall outcome in the first cluster is 0.33 ($SE = 0.15$, $p = 0.01$), and in the second cluster, 0.77 ($SE = 0.15$, $p < 0.001$). Studies with a longer duration have larger effect sizes. In both clusters, the null-hypothesis of homogeneous outcomes is rejected. The proportion of between-study variance is estimated as 0.61, $\chi^2(10)= 22.35$, $p = 0.01$ in the first cluster, and 0.69, $\chi^2(8) = 16.72$, $p = 0.03$ in the second cluster. We can perform a formal test for the difference between the two outcomes. Completely analogous to analysis of variance, where the total variance is partitioned into a between-groups variance and a within-groups variance, we can partition the total chi-square into a between-clusters chi-square and a within-clusters chi-square (Cooper, 1998; Hedges & Olkin, 1985). The total chi-square is the chi-square for the between-study variance for all 20 studies, which is $\chi^2(19) = 48.85$, $p < .001$. The within-clusters chi-square is given by the sum of the chi-squares for the variance within the two clusters, $\chi^2(18) = 39.07$, $p < 0.003$. The between-clusters chi-square is found by subtracting the within-clusters chi-square from the total chi-square; $\chi^2(1) = 9.78$, $p < = 0.002$. The between-clusters chi-square is highly significant ($p < 0.002$). The conclusion seems warranted that duration of experimental intervention has an effect on the outcome. However, the within-clusters chi-square was also significant. The fact that we still have significant heterogeneity in the two clusters indicates that we have not explained all systematic differences between the studies.

MULTILEVEL META-ANALYSIS

A multilevel meta-analysis of the 20 studies using the empty intercept–only model produces virtually the same results as the classical meta-analysis reported earlier. The intercept, which in the absence of other explanatory

TABLE 5.3

Results of Random Effects Model and Multilevel Regression Analyses on
Example Data

Model	Classical Random Effects	Multilevel Null Model	Multilevel Using Duration
intercept	$\delta = 0.58$ (.11)	$\gamma_0 = 0.58$ (.11)	$\gamma_0 = 0.57$ (.08)
duration			$\gamma_1 = 0.14$ (.03)
parameter variance σ_u^2	.17	.14	.04
p value χ^2 test[a]	$p < .001$	$p < .001$	$p = .09$

[a]Chi-square test on residuals, cf. Hedges & Olkin, 1985, and Bryk & Raudenbush, 1992.

variables is the overall outcome that classical meta-analysis indicates by δ, is estimated as $\gamma_0 = 0.58$, with a standard error of 0.11 ($p < 0.001$). The null hypothesis of homogeneous outcomes is rejected, but the between-study variance is estimated a bit lower than in the classical meta-analysis. The between-study variance σ_u^2 is estimated as 0.14, and the proportion of between-study variance is 0.61. This still is much larger than 0.25, the lower limit for examining differences between studies (Hunter & Schmidt, 1990).

The power of multilevel meta-analysis becomes apparent when we attempt to model the differences in the study outcomes. We simply include the duration of the experimental intervention as an explanatory variable in the model. The multilevel meta-analysis model can be written as

$$d_j = \gamma_0 + \gamma_1 Duration_{1j} + u_j + e_j \tag{5.6}$$

The advantage of directly including duration as an explanatory variable is that we do not have to dichotomize or discretize it, as we were forced to do in the clustering approach. The results of the multilevel meta-analysis are summarized in Table 5.3. This table presents the results for both the empty (null) model and the model that includes duration, in addition to the results obtained by the classical (random effects) meta-analysis method.

After including duration as an explanatory variable in the model, the residual between-study variance is no longer significant. The regression coefficient for duration is 0.14 ($p < 0.001$), which means that for each additional week, the expected gain in study outcome is 0.14. The explanatory variable duration is centered on its overall mean, and as a result, the intercept remains essentially unchanged from one model to the next, and it reflects the expected outcome of the average study. The residual variance in the model is 0.04, which is not significant. If we compare this with the between-study variance of 0.14 in the null model, we conclude that 71% of the between-studies variance can be explained by including duration as the explanatory variable in the model.

Because the study outcome depends in part on the duration of the experiment, reporting an overall outcome does not convey all the relevant information. We could report the expected outcomes for different durations (i.e., report the dose-response curve), or calculate which duration is minimally needed to obtain a significant outcome. This is easily done by centering the explanatory variable on different values. For instance, if we center the duration around 2 weeks, the intercept can be interpreted as the expected outcome at 2 weeks. Some multilevel analysis programs can produce predicted values with their expected error variances for various levels of the explanatory variables (such as various durations in this example), which is also useful to describe the expected outcome for experiments with a different duration. Figure 5.1 presents the predicted outcome for our example data for different durations, with the limits of the 95% confidence interval. From the predicted outcome in Fig. 5.1, it is obvious that for low durations negative outcomes are common. Only when the duration of the intervention is at least 4 weeks is the outcome clearly positive.

CORRECTING FOR ARTIFACTS

Hunter and Schmidt (1990, 1994) have advocated to correct study-outcomes for a variety of artifacts. For instance, a common correction is to correct the outcome d for the attenuation that results from unreliability of the measure used. The correction simply divides the outcome measure by the square root of the reliability, for instance $d^* = d/\sqrt{r_{ii}}$, after which the analysis is carried out as usual. This is the same correction as the classical correction for attenuation of the correlation coefficient in psychometric theory (cf. Nunnally & Bernstein, 1994). Hunter and Schmidt (1994) described many other corrections, for instance a correction for the attenuation due to imperfect validity of the outcome measure, corrections for estimated methodological quality, and so on. However, directly applying corrections to the outcome variable results in methodological and statistical problems. A methodological problem is that the majority of these corrections result in larger effect sizes. For instance, if the studies use instruments with a low reliability, the corrected effect size is much larger than the original effect size. Because these large effects have in fact not been observed,

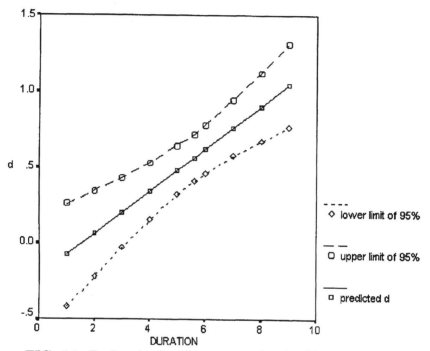

FIG. 5.1. Predicted values for outcome d and 95% interval limits.

automatically carrying out such corrections is controversial. For instance, Schwarzer (1989) advised to always report the original values with the corrected results. A major statistical problem with all these corrections is that their effect on the statistical model is completely unknown. For instance, if the reported reliability is biased, so will be the corrected outcome. If the values used to correct the outcomes are subject to sampling error, and they usually are, the sampling variance of the outcome measure becomes larger. And if many corrections are performed, their cumulative effect on the bias and sampling variance of the outcome measures is totally unclear.

A different and better approach to correct for artifacts is to include them as covariates in the multilevel regression analysis. This is not always optimal; for instance, the attenuation correction follows a multiplicative model, and regression analysis is additive and linear. However, in many cases, adding corrections as explanatory variables to the regression equation produces a reasonable approximation, and when the relationship is not linear, we can always include quadratic or cubic trends in the analysis. For instance, if the range of reliabilities is not extreme, a linear model for the correction is an acceptable approximation. The advantage of this approach is that the effect of measurement unreliability on the study outcomes is estimated based on the available data instead of on a priori corrections. An additional advantage is that we can test whether the correction has a significant contribution to the regression equation. Lastly, if we suspect that a certain covariate has an effect on the variability of the outcomes, we can include it only in the random part of the model, where it affects the between-study variance, but not the average outcome. This models heteroscedasticity at the study level. For instance, it is reasonable to assume that quality of the experimental design used in a study does not necessarily bias the results, but could result in a larger variability of the outcomes. The result would be a larger variance for the residual errors u_j for studies with a poor experimental design. Of course, some covariates might affect both the average outcome and the study-level variance. Models where the residual error variances are a function of other variables were discussed by Goldstein (1995) under the heading "complex variance structures." Although Goldstein did not discuss their application to multilevel data, this is a straightforward extension of his exposition.

A variation on correcting for artifacts is controlling for the effect of sample size. An important problem in meta-analysis is the so-called *file drawer problem*. The data for a meta-analysis are the results from previously published studies. Studies that find significant results may have a larger probability to be published. As a result, a sample of published studies can be biased in the direction of reporting large effects. In classical meta-analysis, one way is to carry out a fail-safe analysis (Greenhouse & Iyengar, 1994). This answers the question of how many unpublished insignificant papers must lie in various researchers' file drawers to render the combined results of the available studies insignificant. If the fail safe number is high, we assume it is unlikely that the file drawer problem affects our analysis. A different approach to the file drawer problem is drawing a *funnel plot*. The funnel plot is a plot of the effect size versus the total sample size (cf. Light & Pillemer, 1984, Light, Singer & Willet, 1994). If the sample of available studies is "well-behaved" this plot should have the shape of a funnel. The outcomes from smaller studies are more variable, but estimate the same underlying population parameter. If large effects are found predominantly in smaller studies, this indicates the possibility of publication bias, and the possibility of many other insignificant small studies remaining unpublished in file drawers.

A problem with the funnel plot is, that we do not know if the smaller studies have different study characteristics too. For instance, small-scale studies could also more often have a short duration. Because the funnel plot is based on observed outcomes, part of the variability in the plot could be due to explanatory study–level variables. In fact, it would be more appropriate to use a funnel plot after removing the covariate effects. An alternative to the funnel plot is to investigate the effect of the study size directly by including the total sample size of a study as explanatory variable in a multilevel meta-analysis. This allows a formal statistical test, and other study characteristics can be controlled simply by adding these to the explanatory variables.

We illustrate those procedures by correcting our example data for reliability of the measure and for total sample size. Table 5.2 has an entry for reliability (r_{ii}). These fictitious data on the effect of social skill training assume that two different instruments were used to measure the outcome of interest; some studies used one instrument, some studies used another instrument. These instruments, in this example, tests for social anxiety in children, differ in their reliability as reported in the test manual. If we use classical psychometric methods to correct for attenuation by unreliability, followed by classical meta-analysis using the random effects model, the combined effect size is estimated as 0.64 instead of the value of 0.58 found earlier. The between-study variance is estimated as 0.23 instead of the earlier value of 0.17. The effect of sample size is more difficult to analyze in classical meta-analysis. A funnel plot indicates a well-behaving sample of studies.

The funnel plot shows virtually no relationship between study outcome and total sample size. We can test this more formally by including total sample size as a covariate in the regression model. If we include the sample size and the reliability as explanatory variables in the regression model, we obtain the results presented in Table 5.4.

The first model in Table 5.4 is the empty intercept–only model presented earlier. Model 2 includes the total sample size, centered on its mean of 54.1, as a predictor, and Model 3 the reliability of the outcome measure, centered on the value 1.0, which represents perfect reliability. Model 4 includes the duration of the experiment, centered on its mean of 5.6. Model 5 includes all available predictors. Both the univariate and the simultaneous analysis show that only duration has a significant effect on the study outcome. Differences in measurement reliability and study size pose no major threats to our substantive conclusion about the effect of duration. If reliability is centered on the value 1, the intercept is estimated as 0.67, which is close to the value of 0.64 estimated in the previous section using the classic attenuation correction on the outcomes. However, the large standard error for the reliability slope suggests that this correction is not necessary. Because there is no relation between the study size and the reported outcome, the existence of a file drawer problem related to sample

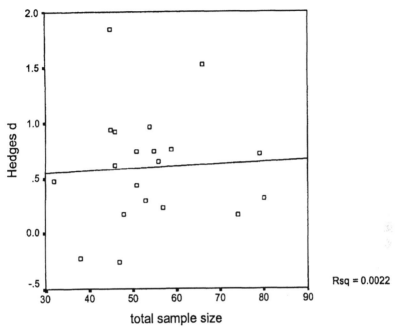

FIG. 5.2. Funnel plot of study outcome against total sample size.

size is unlikely.

The last Model 5 that includes all predictor variables simultaneously is instructive. The (insignificant) regression coefficient for reliability is negative. This is counterintuitive. This is also in the opposite direction of the regression coefficient in Model 3 with reliability as the only predictor. This is a so-called repressor effect caused by the correlations (from 0.25 to 0.33) between the predictor variables. In meta-analysis, because the number of available studies is often small, such effects are likely to occur if we include too many explanatory study-level variables. We conclude that there is an effect of the duration of the treatment on the outcome, and that a bias due to a file drawer problem or differential reliability of the outcome measure is unlikely.

CONCLUSION AND DISCUSSION

The application of multilevel analysis methods in meta-analysis has the advantage that study characteristics can be included in the analysis as potential explanations of the variability of the studies' outcomes. The

TABLE 5.4

Results of Multilevel Regression Including Artifacts as Covariates, all
Covariates Centered

Model	1	2	3	4	5
Predictor:	Intercept only	1 plus N_{tot}	1 plus reliab.	1 plus duration	1 plus all predictors
Intercept	.58 (.11)	.58 (.11)	.67 (.28)	.57 (.08)	.49 (.22)
N_{tot}		.001 (.01)			-.004 (.01)
Reliability			.51 (1.48)		-.52 (1.18)
Duration				.14 (.03)	.15 (.04)
Parameters					
Variance					
σ_u^2	.14	.16	.16	.04	.05
χ^2 test					
p value	$p < .001$	$p < .001$	$p < .001$	$p = .09$	$p = .07$

study characteristics can be theoretically important constructs, or they
can be covariates intended to correct for possible artifacts. Significance
tests of the regression coefficients and predictions can be used to assess the
effect of study characteristics. The residual study–level variance can be
tested for significance to assess whether the study variables explain all the
between–study variance. A comparison of the variance in the empty (null)
model and in the final model informs us how much variance between study
outcomes is explained by our model.

Multilevel regression analysis assumes that all relations are additive
and linear. An additional assumption is that the distribution of the
outcomes is normal (most classical meta-analysis methods require the
same assumption). If these assumptions are violated, transforming the
outcome measure can be helpful (cf. Bryk & Raudenbush, 1992; Hedges
& Olkin, 1985). In practice, we usually have unexplained between-study
variance, as violations of assumptions in the original analyses that produced
the published outcome measures tend to lead to increased variability of
the outcomes. For this reason, we should preferably use models that
include between-studies variance, such as random effects meta-analysis and
multilevel models. Hunter and Schmidt (1994) considered a between-study
variance of up to 25% of the total variance to be uninteresting. In their
view, a larger amount of between study variance cannot be attributed to
various artifacts, and should be further investigated. Following this line of
reasoning, we argue that when we add explanatory variables to a model,

obtaining a residual variance of less than 25% of the total variance is a sign that our model is reasonably complete.

An interesting extension of the multilevel regression model discussed here is allowing for more than two levels. For instance, we may have a situation where there are several outcome measures for each study. The approach in classical meta-analysis is to either combine these into one single outcome per study, or to carry out separate meta-analyses for each different outcome (Gleser & Olkin, 1994). In a multilevel model, it is possible to specify a multivariate outcome model. When all studies report all available outcome measures, the multivariate multilevel model is a straightforward extension of the univariate model (cf. Raudenbush & Bryk, 1985). When some studies do not report on all available outcomes, we have a missing data problem. This extension leads to a more complicated model, which still can be estimated using standard multilevel software. For details, see Kalaian and Raudenbush (1996) and Goldstein (1995). A related extension arises when we have summary data for some studies, whereas we have access to the raw data for others. Goldstein and Yang (2000) showed that such data can be combined in a single model, using standard multilevel analysis software. This allows more refined analyses, using all available data.

The program HLM (Bryk, Raudenbush & Congdon, 1994; Raudenbush, Bryk, Cheong & Congdon, 2000) has a built-in provision for meta-analysis that is restricted to two-levels. If we need three levels, we can use the standard HLM/3L software, using an adapted program setup. The software MLn/MlwiN (Rasbash & Woodhouse, 1995) can also be used for meta-analysis, again with an adapted setup. Ways of tweaking standard multilevel software for meta-analysis are discussed in the Appendix.

There are some minor differences between the programs. HLM uses by default an estimator based on restricted maximum likelihood (RML), whereas MlwiN by default uses full maximum likelihood (FML, called IGLS in MLwiN). Because RML is theoretically better, especially in situations where we have small samples and are interested in the variances, for meta-analysis we should prefer RML (called RIGLS in MlwiN). The results reported earlier were computed using RML. If FML is used, the differences turn out to be small.

An important difference between HLM and MLwiN is the test used to assess the significance of the variances. HLM by default uses a variance test based on a chi-square test of the residuals (Bryk & Raudenbush, 1992). MlwiN estimates a standard error for each variance, which can be used for a Z-test of the variance. In meta-analysis applications, this Z-test is problematic. It is based on the assumption of normality; variances have a chi-square distribution. Especially with small sample sizes and small variances, the Z-test may be inaccurate. An additional advantage of the chi-square test on the residuals is that for the null model it is equivalent to the chi-square variance test in classical meta-analysis (Hedges & Olkin, 1985). The variance tests reported earlier used the chi-square test on the residuals. MLwiN does not offer this test, but it can be produced using the MLwiN macro language.

For estimating complex models, Bayesian procedures are promising and coming into use. These use computer-intensive methods such as Markov Chain Monte Carlo (MCMC) methods to estimate the parameters and their sampling distributions. These methods are attractive for meta-analysis (DuMouchel 1994), because they are less sensitive to the problems that arise when we model small variances in small samples. Bayesian models can be extended by including a prior distribution. This prior distribution reflects a priori beliefs about the likelihood of publication bias. In principle, this is an elegant method to investigate the effect of publication bias. An example of such an analysis is found in Tweedie, Scott, Biggerstaff, and Mengersen (1994). Present multilevel software cannot analyze such models, and more complicated software is needed, such as the general Bayesian modeling program BUGS (Spiegelhalter, 1994).

APPENDIX

Software issues

The simplest program for multilevel meta-analysis is VKHLM, which comes with HLM 2.0 (Bryk et al., 1994) as a separate program, and is built into HLM as an option in later versions. HLM expects for each study a row of data containing a study ID, an outcome measure, its sampling variance, followed by the explanatory variables. If the null model is specified, the results from HLM are close to the classical meta-analysis results produced by Schwarzer's program META, provided one realizes that META reads effect sizes g and transforms these automatically into d's.

Using MLn or MLwiN is more complicated. The data structure is analogous to HLM: We need a study ID, the effect size, its standard error (the square root of the sampling variance), the regression constant (HLM includes this automatically), and the explanatory variables. To set up the analysis, we distinguish two levels: The outcomes are the first level, and the studies the second. Usually we have one outcome per study, so there is no real nesting. The predictor sampling error is included only in the random part on level 1, with a coefficient fixed at 1 (MLwiN uses the command RCON for this). The regression constant is included in the fixed part, and in the random part at level 2. Explanatory variables are included in the fixed part only. MLwiN does not produce the chi-square test on the variances. The formula for the chi-square test is

$$\chi^2 = \sum \left[\left(d_j - \hat{d}_j \right) / SE\left(d_j \right) \right]^2,$$

the sum of the squared residuals divided by their sampling variances. The degrees of freedom are given by $df = J - p - 1$, where J is the number of studies, and p the number of explanatory variables in the model. Assuming that the outcomes are denoted by d, and the standard errors by *sed*, the sequence of MLwiN commands for computing the chi-square is: PRED C50; CALC C50=[(d-C50)/sed]^2; SUM c50 to B1; CPRO B1 *df*. This code assumes that the spreadsheet column C50 is unused.

If we need more than two levels in HLM, we must use HLM/3L, which does not include the VKHLM option. For HLM/3L, we also need a special setup. In this case, we include the standard errors as a weighting variable at the lowest level. We must instruct the program *not* to normalize the weights, which is the default option, and constrain the lowest level variance to be equal to 1.

To apply multilevel models in meta-analysis in other software, this software must have options to set up a model using constraints as specified for MLwiN or for HLM/3L. This means that it must be possible to have a complex lower level variance structure, as in MLwiN, or to constrain the lowest level variance to 1 and to add a weight variable, as in HLM/3L. These options are available in the multilevel options in Lisrel 8.3 (du Toit, du Toit, Jöreskog, & Sörbom, 1999) and in the multilevel analysis program aML (Lillard & Panis, 2000), but these programs do not include the recommended RML estimation. So far, commonly available public domain software for multilevel analysis, such as MixReg (Hedeker & Gibbons, 1996) does not offer the necessary options.

For classical meta-analysis, the program META (Schwarzer, 1989) is freely available from the Internet location http://userpage.fu-berlin.de/health/meta.htm. It comes with a program manual that also explains the basic elements of meta-analysis. The program MetaWin (Rosenberg, Adams & Gurevitch (2000) contains a limited weighted least squares regression option. As indicated earlier, the iterated maximum likelihood methods employed in multilevel analysis is generally more efficient.

References

Bryk, A. S. & Raudenbush, S. W. (1992). *Hierarchical linear models*. Newbury Park, CA: Sage.

Bryk, A. S., Raudenbush, S. W. & Congdon, R. T. (1994). *HLM 2/3. Hierarchical linear modeling with the HLM/2L and HLM/3L programs.* Chicago: Scientific Software International.

Cooper, H. (1998). Synthesizing research: A guide for literature reviews. Thousand Oaks, CA: Sage.

Cornell, J., & Mulrow, C. (1999). Meta-analysis. In H. J. Ader & G. J. Mellenbergh (Eds.), *Research methodology in the social, behavioral, and life sciences,* (pp. 285–323). London: Sage.

DuMouchel, W. H. (1994). *Hierarchical Bayesian linear models for meta-analysis.* Washington, DC: National Institute of Statistical Sciences.

du Toit, S., du Toit, M., Jöreskog, K. G., & Sörbom, D. (1999). *Interactive LISREL user's guide.* Chicago: Scientific Software Inc.

Glass, G. V. (1976). Primary, secondary and meta-analysis of research. *Educational Researcher, 10,* 3–8.

Gleser, L. J., & Olkin, I. (1994). Stochastically dependent effect sizes. In H. Cooper & L. V. Hedges (Eds.), *The handbook of research synthesis* (pp. 339–356). New York: Russel Sage Foundation.

Greenhouse, J. B., & Iyengar, S. (1994). Sensitivity analysis and diagnostics. In H. Cooper & L. V. Hedges (Eds.), *The handbook of research synthesis* (pp. 383–398). New York: Russel Sage Foundation.

Goldstein, H. (1995). *Multilevel statistical models.* London: Edward Arnold.

Goldstein, H., & Yang, M. (2000). Meta-analysis using multilevel models with an application to the study of class size effects. *Applied Statistics, 49* (3), 399–412.

Hedeker, D., & Gibbons, R. D. (1996). MIXREG: A computer program for mixed effects regression analysis with autocorrelated errors. *Computer Methods and Programs in Biomedicine, 49,* 229–252.

Hedges, L. V., & Olkin, I. (1985). *Statistical methods for meta-analysis.* San Diego, CA: Academic Press.

Hedges, L. V., & Vevea, J. L. (1998). Fixed and random effects in meta-analysis. *Psychological Methods, 3* (4), 486–504.

Hox, J. J., 1995. *Applied multilevel analysis (2nd ed.).* Amsterdam: TT-Publikaties.

Hunter, J. E., & Schmidt, F. L. (1990). *Methods of meta-analysis* (pp. 323–336). Newbury Park, CA: Sage.

Hunter, J. E., & Schmidt, F. L. (1994). Correcting for sources of artifact variation. In H. Cooper & L. V. Hedges (Eds.), *The handbook of research synthesis.* New York: Russel Sage Foundation.

Kalaian, H. A., & Raudenbush, S. W. (1996). A multivariate mixed linear model for meta-analysis. *Psychological Methods, 1,* 227–235.

Light, R. J. & Pillemer, D. B. (1984). *Summing up: The science of reviewing research.* Cambridge, MA: Harvard University Press.

Light, R. D., Singer, J. D., & Willet, J. B. (1994). The visual presentation and interpretation of meta-analyses. In H. Cooper & L. V. Hedges (Eds.), *The handbook of research synthesis*(pp. 439–454). New York: Russell Sage Foundation.

Lillard, L. A., & Panis, C. W. A. (2000). *aML. Multilevel multiprocess statistical software, (release 1).* Los Angeles, CA: EconWare.

Nunnally, J. C., & Bernstein, I. H. (1994). *Psychometric theory.* New York: McGraw-Hill.

Overton, R. C. (1998). A comparison of fixed-effects and mixed (random-effects) models for meta-analysis tests of moderator effects. *Psychological Methods, 3* (3), 354–379.

Rasbash, J., & Woodhouse, G. (1995). *MLn command guide.* London: Multilevel Models Project, University of London.

Raudenbush, S. W. (1994). Random effects models. In H. Cooper & L. V. Hedges (Eds.), *The handbook of research synthesis* (pp. 301–322). New York: Russell Sage Foundation.

Raudenbush, S. W., & Bryk, A. S. (1985). Empirical Bayes meta-analysis. *Journal of Educational Statistics, 10,* 75–98.

Raudenbush, S. W., Bryk, A. S., Cheong, Y. F., & Congdon, R. (2000). *HLM5. Hierarchical linear and nonlinear modeling.* Chicago: Scientific Software Inc.

Rosenberg, M. S., Adams, D. C., & Gurevitch, J. (2000). *MetaWin. Statistical software for meta-analysis.* Sunderland, MA: Sinauer Associates.

Rosenthal, R. (1984). *Meta-analytic procedures for social research.* Newbury Park, CA: Sage.

Rosental, R. (1994). Parametric measures of effect size. In H. Cooper & L. V. Hedges (Eds.), *The handbook of research synthesis*(pp. 231–244). New York: Russell Sage Foundation.

Schwarzer, R. (1989). *Meta-analysis programs [computer program manual].* Berlin: Institüt für Psychologie, Freie Universität Berlin.

Snijders, T. A. B., & Bosker, R. (1999). *Multilevel analysis: An introduction to basic and advanced multilevel modeling.* Thousand Oaks, CA: Sage.

Spiegelhalter, D. (1994). *BUGS: Bayesian inference using Gibbs sampling.* Cambridge, England: MRC Biostatistics Unit.

Tweedie, R. L., Scott, D. J., Biggerstaff, B. J., & Mengersen, K. L. (1994). *Bayesian meta-analysis, with application to studies of ETS and lung cancer.* Unpublished report, Colorado State University, Fort Collins.

6

Longitudinal Studies With Intervention and Noncompliance: Estimation of Causal Effects in Growth Mixture Modeling

Booil Jo
Bengt O. Muthén
UCLA

The major interest in intervention trials is often the estimation of intervention effects for individuals who actually receive the intervention. However, some percentage of noncompliance is usually unavoidable in intervention trials when dealing with human participants. In addition, it is not easy to control compliance behavior of individuals who may decide not to participate even with highly attractive incentives. Noncompliance is a major threat to obtaining power to detect intervention effects (Jo, 2002), and may bias the estimation of intervention effects if not handled carefully in the statistical analysis.

ITT (intent to treat) analysis is a standard way to estimate intervention effects in randomized experimental designs in the presence of noncompliance. In this method, average outcomes are compared by randomized groups, ignoring the existence of noncompliance. Because the standard ITT analysis often underestimates intervention effects in the presence of noncompliance, the possibility of estimating intervention effects only for the individuals who actually receive the intervention has been explored using CACE (complier average causal effect) estimation (Angrist, Imbens, & Rubin, 1996; Bloom, 1984; Imbens & Rubin, 1997; Little & Yau, 1998). In CACE approaches, compliers and noncompliers are allowed to be different in various aspects and are best thought of as belonging to different subpopulations. For example, people with higher motivation or a special interest in the intervention will be more likely to comply with the intervention.

In the CACE estimation method, the causal effect of intervention is usually defined based on a single outcome observed after intervention, treating the baseline measure as one of the covariates (i.e., ANCOVA approach). However, when an intervention study is focused on the long-term effect of the intervention, the outcome is often measured several times at specific intervals. In this case, considering the longitudinal nature of intervention studies, it is also possible, and perhaps more natural, to define the intervention effect based on a trend or a growth trajectory of individuals. This study demonstrates CACE estimation based on latent trajectories over time in a growth mixture modeling framework.

One advantage of using a growth modeling framework is that the first time point measure is considered as one of the outcome measures instead of as one of the covariates. This parameterization adds more flexibility in the interpretation of the results because initial status and growth rate of outcome measures are separated. For example, the influence of background variables can be estimated separately for initial status and growth rate of the outcome measure.

Another advantage of this model is that it utilizes not only covariate but also trajectory information, which often improves precision in estimating the compliance type of individuals. Including growth process in the estimation of CACE utilizes the idea of a general latent variable modeling framework where both categorical and continuous latent variables are incorporated (Muthén, 2001a; Muthén, 2001b, Muthén & Muthén, 1998-2001). That is, latent variables that represent growth trajectories are continuous as in conventional structural equation models, whereas the latent variable that represents compliance status is categorical. To differentiate growth modeling with both categorical and continuous latent variables from traditional growth modeling, the former will be called "growth mixture modeling" in this study. This study focuses specifically on random coefficient growth mixture modeling where individual variation is allowed within each class or compliance status. In contrast, individual variation is not allowed within each growth trajectory class in a group-based modeling approach (Nagin, 1999).

This study also explores the possibility of using exploratory growth mixture analysis as a data mining tool that precedes growth mixture CACE analysis. Growth mixture CACE analysis is considered confirmatory because compliance type is known for individuals who are assigned to the intervention condition. In exploratory growth mixture analysis, individuals are classified into several classes without observed class information (training data), and efficiency of classification can often be improved by utilizing the fact that certain trends are present in longitudinal data (Muthén, Brown, Khoo, Yang, & Jo, 1997; Muthén et al., in press; Muthén & Shedden, 1999). When the intervention condition includes many sessions, or doses, one needs to determine the appropriate cutpoints that separate individuals into different classes based on level of compliance. Exploratory growth mixture analysis can be useful in determining cutpoints at the planning stage of growth mixture CACE analysis.

This chapter is organized as follows. Section 1 describes the estimation method using the ML-EM algorithm and defines model assumptions in the estimation of CACE in this study. Section 2 demonstrates the efficiency of CACE estimation in growth mixture modeling through simulation studies. Section 3 demonstrates how exploratory and confirmatory growth mixture analyses can be used in studying unknown subpopulations using the Johns Hopkins Preventive Intervention Study in Baltimore Public Schools as an example. Section 4 concludes with discussion.

ESTIMATING DIFFERENTIAL EFFECTS OF INTERVENTIONS

Model assumptions

The CACE models used in this study are based on statistical assumptions in line with Rubin's causal model. In Rubin's causal model approach, the possibility of statistical causal inference is built based on the effect of treatment assignment at the individual level (Holland, 1986; Rubin, 1974, 1978, 1980). Stable unit treatment value (SUTVA) implies that potential outcomes for each person are unrelated to the treatment status of other individuals (Rubin, 1978, 1980, 1990).

SUTVA and randomization in the study provide a statistical means of causal inference at the population level. Based on these assumptions, four types of subpopulations can be defined by classifying the potential behavior types of the subjects. Angrist et al. (1996) labeled the four categories as complier, never-taker, defier, and always-taker based on assignment and receipt of treatment. Compliers are subjects who do what they are assigned to do. Never-takers are subjects who do not receive the treatment even if they are assigned to the treatment condition. Defiers are the subjects who do the opposite of what they are assigned to do. Always-takers are the subjects who always receive the treatment no matter which condition they are assigned. Among these four kinds of possible compliance types, the current study assumes only two types of compliance and eliminates the possibility of defiers and always-takers. The assumption of monotonicity (Imbens & Angrist, 1994) excludes the possibility of having defiers. In addition, the current study assumes that there are no always-takers, which is the case when study participants are prohibited from receiving a different intervention condition than the one to which they were assigned as in the real data examples shown in later sections. However, unlike monotonicity, the assumption of having no always-takers is not critical in estimating the CACE and can be relaxed depending on the situation.

Unlike ITT analysis, CACE analysis involves methodological complexities due to the missingness of compliance information among control condition individuals. In conventional CACE approaches, it is assumed that the outcome is independent of the treatment assignment for never-takers and always-takers (the exclusion restriction assumption, Angrist et al., 1996). This assumption plays a critical role in simplifying methodological difficulties involved in CACE approaches. Under this assumption, treatment effects are estimated for compliers, but are fixed at zero for the rest. However, this assumption can be unrealistic in some situations (Hirano, Imbens, Rubin, & Zhou, 2000; Jo, in press-a, b). In the Johns Hopkins Preventive Intervention Study example shown in this study, it seems more reasonable to dichotomize individuals as low compliers and high compliers than as never-takers and compliers. In this case, it is possible that the intervention might have a weaker impact on low compliers, but it could not be guaranteed that the intervention has no effect at all, because low compliers were also exposed to the intervention.

Growth mixture CACE modeling using ML-EM estimation method

This study focuses on average causal effect estimation in the random coefficient growth mixture modeling framework. In this study, CACE estimation is used to refer to a more general method that differentiates average causal effect at varying levels of compliance, although the CACE method usually means causal effect estimation that is limited only to compliers.

In growth mixture analysis, the observed outcome variable can be expressed in terms of continuous latent variables that capture growth trajectories over time. Consider a single outcome variable y for individual

i at time point t,

$$y_{it} = I_{ik}\,\lambda_{It} + S_{ik}\,\lambda_{St} + \epsilon_{it}, \qquad (6.1)$$

where latent categorical variable **c** has K levels of compliance status ($k = 1, 2, ..., K$). Compliance status **c** is observed in the intervention group and latent (missing) in the control group. Variable $c_i = (c_{i1}, c_{i2}, ..., c_{ik})$ has a multinomial distribution, where $\mathbf{c}_{ik} = 1$ if individual i belongs to class k and zero otherwise. The categorical latent variable approach may also be referred to as finite mixture modeling where sampling units consist of subpopulations that might have separate distributions and different model parameters (McLachlan & Peel, 2000; Titterington, Smith, & Makov, 1985). In finite mixture modeling, the number of mixture components is assumed to be known and fixed. For example, $K = 2$ in simulation studies and real data examples shown in later sections. Here, I_{ik} and S_{ik} are individually varying continuous latent variables representing initial level of outcome and growth rate (slope) respectively. The time scores for the initial status (λ_{It}) are equal across all time points (usually fixed at 1.0) because initial status does not change over time. The time scores for the growth rate (λ_{St}) are 0, 1, 2,..., T, representing linear growth over time, which may be fixed at different values depending on the distance between the measuring points. And ϵ_{it} represents a normally distributed residual at time point t with zero mean and variance σ_t^2.

Individual variation in growth parameters I_{ik} and S_{ik} within compliance class k can be expressed as

$$I_{ik} = I_k + \gamma_{Ix}\,\mathbf{x}_i + \zeta_{Iik}, \qquad (6.2)$$
$$S_{ik} = S_k + \gamma_{Sx}\,\mathbf{x}_i + \gamma_{Zk}\,Z_i + \zeta_{Sik}, \qquad (6.3)$$

where I_k and S_k represent intercept parameters of initial status and slope for each compliance class k; **x** represents a vector of observed covariates, and γ_{Ix} and γ_{Sx} are regression coefficient parameters. And ζ_{Iik} and ζ_{Sik} are normally distributed residuals with zero means and variances ψ_{Ik}, ψ_{Sk}, and a covariance ψ_{ISk}. The Z_i is a binary variable that represents intervention assignment, where $Z_i = 1$ if individual i is assigned to the intervention condition and zero if individual i is assigned to the control condition. Based on randomization, growth rate (slope) is regressed on Z_i, but initial status is not regressed on Z_i. The γ_{Zk} represents a mean shift in the slope when subject i belongs to the intervention condition, and is allowed to vary across different compliance status. In this study, intervention effect is defined as the difference between intervention and control conditions in the outcome measure at the final time point. Based on Equations 6.1, 6.2 and 6.3, the average causal effect (ACE) of an intervention assignment can be defined at compliance level k at the last time point (i.e., $\lambda_{St} = T$) as

$$ACE_k = \gamma_{Zk} \times T. \tag{6.4}$$

The class probability $\boldsymbol{\pi}_i$ is allowed to vary as a function of covariates. When background variables are available, the multinomial logit model of $\boldsymbol{\pi}_i$ with a vector of covariates $\mathbf{x}$ is decribed as

$$logit(\boldsymbol{\pi}_i) = \boldsymbol{\beta}_0 + \boldsymbol{\beta}_1 \mathbf{x}_i, \tag{6.5}$$

where $\boldsymbol{\pi}_i$ is a K dimensional vector $(\pi_{i1}, \pi_{i2}, ..., \pi_{iK})'$, $\pi_{ik} = P(c_{ik} = 1 \mid \mathbf{x}_i)$, and $logit(\boldsymbol{\pi}_i) = [log(\pi_{i1}/\pi_{iK}), log(\pi_{i2}/\pi_{iK}), ..., log(\pi_{i,K-1}/\pi_{iK})]'$. The $\boldsymbol{\beta}_0$ are $K-1$ dimensional logit intercepts, and $\boldsymbol{\beta}_1$ are multinomial logit regression coefficient parameters. The multinomial logit regression also provides information about the characteristics of individuals with different compliance levels.

CACE analyses reported in this study were carried out by the M*plus* program (Muthén & Muthén, 1998-2001) using maximum likelihood estimation via the EM algorithm (Dempster, Laird, & Rubin, 1977; Little & Rubin, 1987; McLachlan & Krishnan, 1997; Tanner, 1996). In the ML-EM method, the unknown compliance status (c) is handled as missing data.

Consider the sampling distribution of $\mathbf{y}$ and $\mathbf{x}$ from the mixture of k components

$$g(\mathbf{y}, \mathbf{x} \mid \boldsymbol{\theta}, \boldsymbol{\pi}) = \sum_{k=1}^{K} \pi_k f(\mathbf{y}, \mathbf{x} \mid \boldsymbol{\theta}_k), \tag{6.6}$$

where $\mathbf{y}$ and $\mathbf{x}$ represent observed data, $\boldsymbol{\theta}$ represents model parameters, and π_k represents the proportion of the population from component k with $\sum_{k=1}^{K} \pi_k = 1$. The probability $\boldsymbol{\pi}$ is the parameter that determines the distribution of $\mathbf{c}$. The observed-data log likelihood is

$$LogL = \sum_{i=1}^{n} log(\mathbf{y}_i \mid \mathbf{x}_i). \tag{6.7}$$

Given the formulation of the proposed growth mixture CACE model, the complete-data log likelihood can be written as

$$LogL_c = \sum_{i=1}^{n} [log(\mathbf{c}_i \mid \mathbf{x}_i) + log(\boldsymbol{\eta}_i \mid \mathbf{c}_i, \mathbf{x}_i) + log(\mathbf{y}_i \mid \boldsymbol{\eta}_i)], \tag{6.8}$$

where

$$\sum_{i=1}^{n} log(\mathbf{c}_i \mid \mathbf{x}_i) = \sum_{i=1}^{n} \sum_{k=1}^{K} c_{ik} \, log \, \pi_{ik}. \tag{6.9}$$

In Equations 6.8 and 6.9, $\mathbf{c}$ represents categorical latent compliance class, and $\boldsymbol{\eta}$ represents continuous latent growth factors (i.e., I and S).

Maximum likelihood estimation using the EM algorithm considers complete-data log likelihood shown in Equation 6.8. The E step computes the expected values of the complete data–sufficient statistics, given data and current parameter estimates. Compliance status $\mathbf{c}$ is considered as missing data in this step. The conditional distribution of $\mathbf{c}$, given the observed data and the current value of model parameter estimates $\boldsymbol{\theta}$, is given by

$$f(\mathbf{c} \mid \mathbf{y}, \mathbf{x}, \boldsymbol{\theta}) = \prod_{i=1}^{n} f(\mathbf{c}_i \mid \mathbf{y}_i, \mathbf{x}_i, \boldsymbol{\theta}). \tag{6.10}$$

The E step applies to both confirmatory (i.e., CACE) and exploratory growth mixture analyses, but the difference is that growth mixture CACE analysis uses information about already–known class membership (i.e., compliance status) in the intervention condition. Therefore, the first step of the implentation is easily modified with a known value of the indicator c_{ik}.

The M step computes the complete data ML estimates with complete data–sufficient statistics replaced by their estimates from the E step. This procedure continues until it reaches optimal status. The M step maximizes

$$\sum_{i=1}^{n} \sum_{k=1}^{K} p_{ik} \, log \, \pi_{ik} \tag{6.11}$$

with respect to model parameters. The p_{ik} is the posterior class probability of individual i, conditioning on observed data and model parameters, where $\pi_{ik} = P(c_{ik} \mid \mathbf{x}_i)$.

The identifiability and precision of mixture and growth mixture models used for CACE analyses in this study are based on observed compliance class membership in the intervention condition (training data) and various sources of auxilliary information such as from covariates and growth trajectories. For more details about identifiability and efficiency of extended CACE models, see Jo (in press-a). Parametric standard errors are computed from the information matrix of the ML estimator using both the first– and the second–order derivatives under the assumption of normally distributed outcomes. For more details about estimation procedures in growth mixture modeling, see Muthén & Muthén (1998-2001), Muthén & Shedden (1999), and Muthén et al. (in press).

CACE ESTIMATION USING THE OBSERVED AND LATENT VARIABLES: SIMULATION STUDIES

In longitudinal intervention studies, the effect of the intervention can be defined as the difference between the intervention and the control group in the observed outcome measured at the last time point, conditioning on the outcome measured at the first time point (ANCOVA approach). An alternative is to define the intervention effect based on latent variables that capture growth trajectories of individuals (growth model approach) as described in the previous section. This section demonstrates the quality of average causal effect estimates based on observed variable (ANCOVA) and latent variable (growth model) approaches. The simulation studies shown in this section assume that there are two underlying subpopulations with different compliance behaviors ($K = 2$). One subpopulation consists of individuals who would show a high level of compliance if assigned to the intervention condition (high compliers). The other subpopulation consists of individuals who would show a low level of compliance if assigned to the intervention condition (low compliers). The ratio of high and low compliers is 50:50, and the ratio of individuals assigned to the intervention and control conditions is 50:50. It is assumed that the intervention assignment is binary (intervention condition if $Z = 1$, control condition if $Z = 0$) and has differential effects on high compliers and low compliers. The true parameter values, effect size, and sample size are chosen based on the Johns Hopkins Public School Preventive Intervention Study example that is shown in a later section. Covariates are not included in this setting.

The true initial status mean (I_h) and the true mean growth rate (S_h) for high compliers are

$$\begin{pmatrix} I_h \\ S_h \end{pmatrix} = \begin{pmatrix} 5.00 \\ -0.25 \end{pmatrix}.$$

The true initial status mean (I_l) and the true mean growth rate (S_l) for low compliers are

$$\begin{pmatrix} I_l \\ S_l \end{pmatrix} = \begin{pmatrix} 3.50 \\ 0.15 \end{pmatrix}.$$

The true additional growth rates for high compliers (γ_{Zh}) and low compliers (γ_{Zl}) when they are assigned to the intervention condition are $\gamma_{Zh} = 0.20$, $\gamma_{Zl} = -0.10$, where the positive value of γ_{Zh} represents a desirable effect of the intervention for high compliers, and the negative value of γ_{Zl} represents a negative effect of the intervention for low compliers, assuming that positive growth of the outcome is desirable.

The true initial status variance is the same for high compliers and low compliers ($\psi_{Ih} = \psi_{Il} = 0.64$), and the true growth rate residual variance is the same for high compliers and low compliers ($\psi_{Sh} = \psi_{Sl} = 0.0625$). The true residual covariance between initial status and growth rate is zero for both high compliers and low compliers ($\psi_{ISh} = \psi_{ISl} = 0$), but is not fixed at zero in the analyses. Both variances and covariances are assumed to be equal across high and low compliers in the analyses. However, in real data examples shown in a later section, both variances and covariances are allowed to vary across high and low compliers.

It is assumed in the simulation setting that the outcome is measured four times with equal distances and has a linear trend over time. The initial status does not change over time. Given that, the fixed time scores for initial status and growth rate used in both data generation and growth mixture CACE analyses are

$$
(\lambda_{It}, \ \lambda_{St}) = \begin{pmatrix} 1 & 0 \\ 1 & 1 \\ 1 & 2 \\ 1 & 3 \end{pmatrix}.
$$

The true residual variances and covariances of observed outcome measures are

$$
\begin{pmatrix} \sigma_{11} & \sigma_{12} & \sigma_{13} & \sigma_{14} \\ \sigma_{21} & \sigma_{22} & \sigma_{23} & \sigma_{24} \\ \sigma_{31} & \sigma_{32} & \sigma_{33} & \sigma_{34} \\ \sigma_{41} & \sigma_{42} & \sigma_{43} & \sigma_{44} \end{pmatrix} = \begin{pmatrix} 0.36 & 0 & 0 & 0 \\ 0 & 0.49 & 0 & 0 \\ 0 & 0 & 0.64 & 0 \\ 0 & 0 & 0 & 1.00 \end{pmatrix},
$$

where true residual covariances are zero, and are also assumed to be zero in the analyses. The true residual variances of outcome measures result in R^2 of 0.64 at the first time point, and 0.55 at the last time point.

The model used for data generation and the CACE analysis using the latent variable approach can be described as

$$
\begin{aligned}
y_{it} &= I_{ik} \lambda_{It} + S_{ik} \lambda_{St} + \epsilon_{it}, & (6.12) \\
I_{ik} &= I_k + \zeta_{Iik}, & (6.13) \\
S_{ik} &= S_k + \gamma_{Zk} Z_i + \zeta_{Sik}, & (6.14)
\end{aligned}
$$

where the assignment of an intervention has a differential effect (γ_{Zk}) on the growth rate of high compliers and low compliers. According to Equation 6.4, γ_{Zk} can be translated into the intervention effect at the last time point (i.e., $\gamma_{Zk} \times 3$).

Given that $K = 2$ and there are no covariates, the multinomial logit model can be simplified as

$$P(i \in h) = \pi_{hi},$$
$$P(i \in l) = 1 - \pi_{hi} = \pi_{li},$$
$$logit(\pi_{hi}) = \beta_0, \tag{6.15}$$

where π_{hi} denotes the probability of being a high complier, π_{li} denotes the probability of being a low complier, and β_0 represents a logit intercept that determines the ratio of high and low compliers. The true logit intercept is 0.0 (i.e., 50:50).

The model used for the CACE analysis based only on the observed variables (ANCOVA approach) can be expressed as

$$y_{i4} = \alpha_k + \lambda_x \, y_{i1} + \Gamma_{Zk} \, Z_i + \epsilon_{i4}, \tag{6.16}$$

where the differential effect of intervention (Γ_{Zk}) is defined based on the outcome measured at the last time point (y_{i4}) conditioning on the outcome measured at the first time point (y_{i1}). Here, the baseline outcome measure (y_{i1}) is considered as a covariate.

In the ANCOVA approach, the baseline outcome measure (y_{i1}) is also used as a predictor of compliance. Treating y_{i1} as a covariate, the logit model can be described as

$$P(i \in h \mid y_{i1}) = \pi_{hi},$$
$$P(i \in l \mid y_{i1}) = 1 - \pi_{hi} = \pi_{li},$$
$$logit(\pi_{hi}) = \beta_0 + \beta_1 \, y_{1i}, \tag{6.17}$$

where the logit coefficient β_1 shows the level of association between the baseline outcome measure and compliance behavior.

The simulation results presented in Table 6.1 are based on 500 replications with a sample size of 300. Coverage is defined as the proportion of replications out of 500 replications where the true intervention effects for high and low compliers are covered by the 95% confidence intervals of intervention effect estimates. Power is defined as the proportion of replications out of 500 replications where the intervention effect estimate is significantly different from zero ($\alpha = .05$).

It is demonstrated in Table 6.1 that both the ANCOVA and growth model approaches provide average intervention effect estimates with reasonable quality, considering that compliance information is missing for 50% of individuals (i.e., control condition individuals) and the sample size is fairly small (i.e., $N = 300$). Simulation results show that the quality

TABLE 6.1

CACE Analyses Using the Observed and Latent Variable Approaches: The Quality of Average Intervention Effect Estimates at the Last Time Point

	Observed Variable (ANCOVA) Approach		
Intervention Effect	True Value	Avg Estimate	Avg SE
High Complier (Γ_{Zh})	0.60	0.604	0.292
Low Complier (Γ_{Zl})	-0.30	-0.280	0.291
Coverage (High Complier)		**0.930**	
Coverage (Low Complier)		**0.932**	
Power (High Complier)		**0.560**	
Power (Low Complier)		**0.226**	
	Latent Variable (Growth Model) Approach		
Intervention Effect	True Value	Avg Estimate	Avg SE
High Complier ($\gamma_{Zh} \times 3$)	0.60	0.606	0.259
Low Complier ($\gamma_{Zl} \times 3$)	-0.30	-0.286	0.260
Coverage (High Complier)		**0.940**	
Coverage (Low Complier)		**0.938**	
Power (High Complier)		**0.657**	
Power (Low Complier)		**0.248**	

of estimates is close between the two models in terms of point estimates and standard errors, implying comparability of the two models. The growth model approach, however, shows slightly better point estimates and standard errors than the ANCOVA approach. Although the gap between the two approaches in point estimates and standard errors is not dramatic, it still results in a noticeable difference between the two methods in terms of statistical power to detect intervention effects (e.g., 0.657 vs. 0.560 for high compliers). In the ANCOVA approach, only the outcome measures at the first (y_1) and the last time point (y_4) are considered, and the outcome measures in between (y_2, y_3) are ignored. The loss in information may lead to a lower precision in the ANCOVA approach.

It has been demonstrated in previous research that average causal effects of interventions can be identified for more than one subpopulation with a satisfactory level of accuracy based on auxiliary information from covariates (Jo, in press-a). Simulation results shown in Table 6.1 show that growth trajectories (in a latent variable form) can provide auxiliary information that can be used for the same purpose. It is also shown that the growth model approach may improve precision of average causal effect estimates by handling measurement errors and by utilizing trajectory information.

THE JOHNS HOPKINS PUBLIC SCHOOL PREVENTIVE INTERVENTION STUDY

The Johns Hopkins Public School Preventive Intervention Study was conducted by the Johns Hopkins University Preventive Intervention Research Center (JHU PIRC) in 1993 to 1994 (Ialongo et al., 1999). Based on the life course/social field framework as described by Kellam and Rebok (1992), the Johns Hopkins PIRC preventive trial focused on successful adaptation to first grade as a means of improving social adaptational status over the life course. The study was designed to improve academic achievement and to reduce early behavioral problems of school children. Teachers and first-grade children were randomly assigned to intervention conditions. The control condition and the family-school partnership intervention condition are compared in this example. In the intervention condition, parents were asked to implement 66 take-home activities related to literacy and mathematics over a 6 month period. The intervention was provided over the first-grade school year (1993–1994), following a pretest assessment in the early fall. The intervention impact was assessed in the spring of first (6 months from the pretest) and second (18 months from the pretest) grades. In the spring of first grade, 91.3% completed assessments, and in the spring of second grade, 88.5% completed assessments.

A total sample size of 333 was analyzed after listwise deletion of cases that had missingness in covariates and outcome variables. The two major outcome measures in the JHU PIRC preventive trial were academic achievement (CTBS mathematics and reading test scores) and the TOCA-R score (Teacher Observation of Classroom Adaptation-Revised; Werthamer-Larsson, Kellam, & Wheeler, 1991). The TOCA-R is designed to assess each child's adequacy of performance on the core tasks in the classroom as rated by the teacher. Among various outcome measures, readiness to learn (or work) assessed in the spring of the second grade (18 months from the pretest) is used as the outcome in this example. In the JHU PIRC preventive trial, readiness to learn does not represent acquisition of prerequisite knowledge or skills, but rather, being ready to exert effort to reach academic excellence. The readiness to learn scale ranges from 1 to 6, and consists of TOCA-R items that measure whether a child completes assignments, puts forth effort, and works hard. Table 6.2 shows the sample statistics for the variables used in the analyses of this study.

Intent to treat analysis using the ANCOVA approach

Standard ITT analysis provides an overall average intervention effect estimate by comparing the outcome based on assignment of intervention, but ignoring the aspect of the receipt of the intervention. That is, it assumes that children of parents with a low compliance rate receive the same effects from the intervention as children of parents with a high compliance rate. Table 6.3 shows the results from the JHU PIRC preventive trial data analysis using the ITT analysis. In this analysis, the overall effect of the intervention is estimated based on the outcome measured at the last time point (Ready3). The outcome measured at the first time point (Ready1) is used as one of the covariates, and the outcome measured at the second time point (Ready2) is not considered in the analysis (ANCOVA approach).

There is a positive effect of the intervention on the level of children's readiness to learn (intervention effect = 0.316, effect size = 0.212). The effect size of the intervention was calculated by dividing the outcome difference in the intervention and the control condition means by the square root of the variance pooled across the control and intervention groups. In the ITT analysis, baseline readiness to learn (Ready1) and free lunch program were found to be significant predictors of the level of readiness to learn. Children had a higher level of readiness at the last time point if their baseline readiness level was higher, and a lower level of readiness if their SES background level was low.

TABLE 6.2

Johns Hopkins PIRC: Sample Statistics (N=333)

Variable	Mean	SD	Description
Z	0.52	0.50	Intervention assignment (0 = control, 1 = intervention)
Ready1	4.59	1.32	TOCA mean readiness at the pretest
Ready2	4.48	1.39	TOCA mean readiness 6 months from the pretest
Ready3	4.33	1.49	TOCA mean readiness 18 months from the pretest
Male	0.49	0.50	Student's gender (0 = female, 1 = male)
Lunch	0.60	0.49	Free lunch program (0 = no, 1 = yes)
Unemployed	0.14	0.34	Parent's employment status (0 = no, 1 = yes)
Married	0.47	0.50	Parent's marital status (0 = no, 1 = yes)
Limited Health	0.10	0.30	Parent limited by health problem (0 = no, 1 = yes)
Health	3.83	1.03	Parent's overall health (1 = poor, 5 = excellent)
Age	2.97	1.42	Parent's age in 5-year brackets

TABLE 6.3

Johns Hopkins PIRC: Intent to Treat Analysis

Parameter	Estimate	SE
Intervention effect	0.316	0.147
Ready3 Regressed on Covariates		
Ready1	0.505	0.060
Male	-0.243	0.143
Free Lunch	-0.410	0.156
Unemployed	-0.348	0.236
Married	-0.288	0.147
Limited Health	-0.074	0.236
Health	-0.062	0.077
Age	-0.037	0.050
Intercept	2.742	0.503
$\sigma_3{}^2$	1.636	0.112

CACE analysis using the ANCOVA approach

In the ITT analysis, intervention effect may be underestimated for high compliers due to the inclusion of low compliers who might not have been exposed enough to benefit from the intervention. In this situation, the possible bias can be avoided by taking into account the difference between the two subpopulations in the analysis. Table 6.4 shows the results from the CACE analysis, where the differential effect of intervention is estimated for high compliers and low compliers. As in the ITT analysis, intervention effect is estimated based only on observed variables (ANCOVA approach). The same set of covariates used in the ITT analysis are used as predictors of the outcome, and also as predictors of compliance. The model used for this CACE analysis is the same as the model described in Equations 6.16 and 6.17, with the exception that more covariates are included in this example in addition to the baseline outcome measure.

Table 6.4 shows that the intervention had a positive impact on the

TABLE 6.4

Johns Hopkins PIRC: CACE Analysis Using the Observed Variable (ANCOVA) Approach

Parameter	Estimate	SE
Intervention Effect		
High Complier (Γ_{Zh})	0.477	0.221
Low Complier (Γ_{Zl})	-0.103	0.530
Ready3 Regressed on Covariates		
Ready0	0.524	0.060
Male	-0.246	0.145
Free Lunch	-0.419	0.157
Unemployed	-0.355	0.238
Married	-0.276	0.147
Limited Health	-0.078	0.243
Health	-0.059	0.079
Age	-0.048	0.051
High Comp Intercept (α_h)	2.523	0.523
Low Comp Intercept (α_l)	3.093	0.723
σ_3^2	1.606	0.118
c Regressed on Covariates (High vs. Low Compliers)		
Ready0	0.376	0.130
Male	-0.021	0.350
Free Lunch	-0.235	0.385
Unemployed	-0.162	0.492
Married	0.095	0.347
Limited Health	-0.266	0.720
Health	0.009	0.191
Age	-0.221	0.109
Logit Intercept (β_0)	0.059	1.116

level of readiness for children with parents with a high compliance rate (intervention effect = 0.477, effect size = 0.320), and the magnitude of the effect was larger than that of the overall effect in the ITT method. It also shows a slightly negative effect of the intervention for children of parents with a low compliance rate, but the magnitude of the effect is very small and insignificant (intervention effect = -0.103, effect size = -0.069). Effect size was calculated based on a pooled standard deviation as in the ITT analysis. This approach was chosen for easier comparison across different estimation methods. In this analysis, baseline readiness to learn (Ready1) and free lunch program were found to be significant predictors of the level of readiness to learn.   Children had a higher level of readiness at the last time point if their baseline readiness level was higher, and a lower level of readiness if their SES background level was low. Initial level of child's readiness and parent's age were found to be significant predictors of parent's compliance behavior. Parents complied more if the child's baseline readiness level was higher. Younger parents also complied more.

For the CACE analysis shown in Table 6.4, individuals were dichotomized into either the low or the high complier category based on the level of completeness in home learning activities. For easier comparison, the same cutpoint is used as that in the CACE analysis using the latent variable approach that is shown in a later section. For illustration purposes, compliance was dichotomized in this example; but note that sensitivity of the CACE estimate to different thresholds needs to be carefully examined in practice (West & Sagarin, 2000). The following section shows how the cutpoint was decided in this study for CACE analyses.

Exploratory growth mixture analysis

This section examines the possibility of using exploratory growth mixture analysis as a data-mining tool that precedes CACE analysis using mixture and growth mixture models.   In randomized intervention trials, the intervention condition often includes many sessions, or doses. One way to model compliance behavior in this situation is to treat compliance as a continuous variable.   Holland (1988) proposed ALICE (additive linearly constant effects) model, where the effect of intervention is estimated based on continuous compliance.   The ALICE model requires several strong assumptions, which often limits the applicability of the model in practice. For example, it is assumed in the ALICE model that the effect of intervention linearly increases as the level of compliance increases. In the JHU PIRC preventive trial, there is a large variation in completed number of intervention activities (range 0 to 66), and children may not get any benefit from the intervention unless parents complete a sufficient number of activities. Over reporting of compliance level is also expected, because parents self-report their level of completion in intervention activities. In this situation, the intervention may not show any desirable effects unless parents report a quite high level of compliance. Another way to model compliance behavior in this situation is to use the dose-response curve approach (Efron & Feldman, 1991), where the effect of intervention can be estimated without assuming a linear relationship between intervention effects and compliance. However, this approach requires successful double-blind experiments, which are not often applicable especially in psychosocial intervention trials. The third way to model compliance behavior in this situation is to treat compliance as a categorical variable without assuming linearity. The difficulty of this approach is in deciding the appropriate number of categories and thresholds that separate individuals into different compliance categories.

The current study takes the third approach in analyzing the Johns Hopkins PIRC preventive trial data, and shows that exploratory growth mixture analysis can be useful in determining cutpoints at the planning stage of CACE analysis. To estimate the differential effect of the intervention for those who completed enough activities and for those who did not, the compliance measure is dichotomized in this example. Exploratory growth mixture analysis is conducted for control group individuals, which provides the information about the trajectory shape and the proportion of subgroups in the absence of intervention (Muthén et al., in press). The confirmatory mixture analysis (i.e., CACE analysis) following the exploratory analysis is based on the idea that subpopulations that are already different in the absence of intervention will be more likely to differ in terms of compliance behavior. Consequently, it is also expected that the effect of intervention will differ for these heterogeneous subpopulations.

The model used for exploratory growth mixture analysis is the same as the model described in Equations 6.1, 6.2, 6.3, and 6.5 except that $\gamma_{Zk} Z_i$ is removed from Equation 6.2 in the exploratory analysis. The same covariates used in CACE analyses in the previous section and in the following section are used in the exploratory growth mixture analysis. However, note that the model used for exploratory analysis is significantly different from the model used for mixture and growth mixture CACE analyses because the intervention group is not included in the model and potential level of compliance is not considered in estimating class membership of individuals. Figure 6.1 shows estimated trajectories suggested by two-class exploratory growth mixture analysis for the control group.

Figure 6.1 shows that the level of readiness decreases over time for the majority of children (69.2%), whose baseline readiness level is high. It also shows that the level of readiness increases over time for the other class

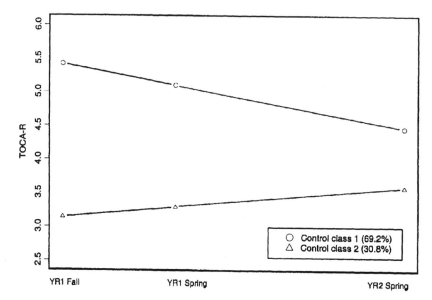

FIG. 6.1. Estimated mean curves of readiness to learn in the control group using exploratory growth mixture analysis.

of children (30.8%), whose baseline readiness level is low. Based on the proportion of subpopulations from the exploratory analysis, parents who completed 35 or more of the take-home learning activities were categorized as high compliers (71% of parents), and parents who completed fewer than 35 take-home learning activities were categorized as low compliers (29% of parents). Four parents did not comply at all and were included in the low complier category in this example. Parents in the control condition could not be dichotomized because their compliance information was missing. Figure 6.2 shows observed mean curves of readiness to learn based on this dichotomization.

CACE analysis using the growth model approach

This section demonstrates the estimation of intervention effects using the growth mixture modeling approach, where the effect of intervention is defined based on a trend or a growth trajectory of individuals. The growth mixture model used for CACE analysis is the same as the model described in Equations 6.1, 6.2, 6.3, and 6.5. The same covariates used in the CACE analysis using the ANCOVA approach and the exploratory growth mixture analysis are used. Based on exploratory growth mixture analysis of the control group and observed mean curves shown in Figure 6.2, linear

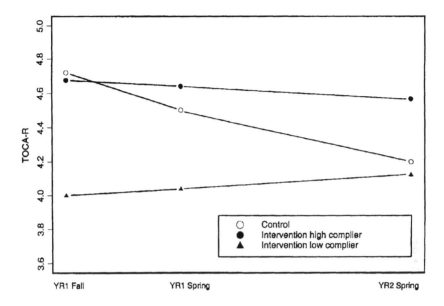

FIG. 6.2. Observed mean curves of readiness to learn.

trajectory was found to be appropriate for the CACE analysis using growth model approach. In the JHU PIRC preventive trial example, the outcome was measured in fall of the first grade, spring of the first grade, and spring of the second grade (Ready1, Ready2, Ready3). Given the distances between time points, the time scores used to capture a linear trend over time are 0, 1, and 3. Therefore, the average causal effect of intervention assignment at the last time point is defined as $\gamma_{Zh} \times 3$ for high compliers and $\gamma_{Zl} \times 3$ for low compliers (see Equation 6.4). Table 6.5 shows the results from the CACE analysis using the growth model approach.

Table 6.5 shows that the intervention had a positive impact on readiness of children if parents showed a high compliance rate (intervention effect = 0.477, effect size = 0.320). For children with highly complying parents, the level of readiness to learn decreases significantly less compared to that of control condition children with parents who could have been high compliers if they have had been assigned to the intervention condition. The intervention effect for high compliers in the CACE analysis using the growth model approach has the same magnitude as in the CACE analysis using the ANCOVA approach (see Table 6.4), but the confidence interval is slightly tighter than in the CACE analysis using the ANCOVA approach. Table 6.5 also shows a slightly negative but insignificant effect of the intervention for children of parents with a low compliance rate (intervention effect =

TABLE 6.5

Johns Hopkins PIRC: CACE Analysis Using the Latent Variable (Growth Model) Approach

Parameter	Estimate	SE
Intervention Effect		
High Complier ($\gamma_{Zh} \times 3$)	0.477	0.186
Low Complier ($\gamma_{Zl} \times 3$)	-0.150	0.393
Initial Status Regressed on Covariates		
Male	-0.259	0.134
Free Lunch	-0.244	0.146
Unemployed	-0.068	0.200
Married	0.076	0.138
Limited Health	0.368	0.254
Health	0.041	0.078
Age	0.058	0.047
High Comp Intercept (I_h)	4.682	0.420
Low Comp Intercept (I_l)	3.908	0.427
ψ_{Ih}	0.936	0.139
ψ_{Il}	1.608	0.253
Growth Rate Regressed on Covariates		
Male	-0.033	0.054
Free Lunch	-0.092	0.057
Unemployed	-0.108	0.091
Married	-0.105	0.055
Limited Health	-0.093	0.101
Health	-0.025	0.031
Age	-0.026	0.019
High Comp Intercept (S_h)	0.118	0.158
Low Comp Intercept (S_l)	0.424	0.208
ψ_{Sh}	0.064	0.037
ψ_{Sl}	0.151	0.055
ψ_{ISh}	-0.048	0.051
ψ_{ISl}	-0.227	0.106
σ_1^2	0.444	0.118
σ_2^2	0.708	0.080
σ_3^2	0.717	0.216
c Regressed on Covariates (High vs. Low Compliers)		
Male	-0.175	0.324
Free Lunch	-0.388	0.352
Unemployed	-0.217	0.457
Married	0.102	0.331
Limited Health	0.042	0.621
Health	0.029	0.171
Age	-0.205	0.106
Logit Intercept (β_0)	1.774	0.874

-0.150, effect size = -0.101).

The parameterization used in the growth model approach adds more flexibility in the interpretation of the CACE analysis than that used in the ANCOVA approach because initial status (I) and growth rate (S) are separated. For example, the influence of background variables can be estimated separately for initial level of readiness and change of readiness. In the CACE analysis using the ANCOVA approach, child's gender and free lunch program were found to be significant predictors of the outcome. However, these variables are not significant predictors when initial status (I) and growth rate (S) are separated as shown in Table 6.5. It also shows in CACE analysis using the growth model approach that low compliers have more variation in initial status and growth rate conditioning on covariates. In the high compliance category, initial status and growth rate show very low correlation conditioning on covariates ($\psi_{ISh} = -0.048$). However, in the low compliance category, initial status and growth rate are negatively correlated conditioning on covariates ($\psi_{ISl} = -0.227$). In addition to flexibility in modeling, another advantage of the growth model approach is that it utilizes not only covariates but also trajectory information to identify class membership and to increase efficiency in estimating the differential effect of intervention.

Figure 6.3 shows estimated mean readiness curves over time based on results in Table 6.5. Estimated mean outcomes can be calculated using Equations 6.1, 6.2 and 6.3 and weighted covariate means based on posterior class probability of each individual. This figure shows how readiness to learn changed over time depending on parents' compliance level and intervention assignment. It shows that highly complying parents' children had a higher level of readiness at the first grade, but the level could decrease to a point even lower than that of less involved parents' children by the second grade unless the intervention was given.

By comparing the mean trajectories of the control group in Fig. 6.3 to those in Fig. 6.1, it can be learned how closely subpopulations derived by exploratory and confirmatory growth mixture analyses are related. Mean trajectories in Figs. 6.1 and 6.3 show similarity in the sense that the level of readiness decreases over time for the majority of children (those with high baseline readiness), and the level of readiness increases over time for the other class of children (those with low baseline readiness). Mean trajectories in Figs. 6.1 and 6.3 also show discrepancy in the sense that trajectories in Fig. 6.1 have a larger difference at the initial point and a smaller difference at the last time point than those in Fig. 6.3. The disagreement is not surprising because the model used for exploratory analysis does not consider potential level of compliance, whereas CACE (confirmatory) analysis does. However, information from exploratory analysis is still valuable in deciding

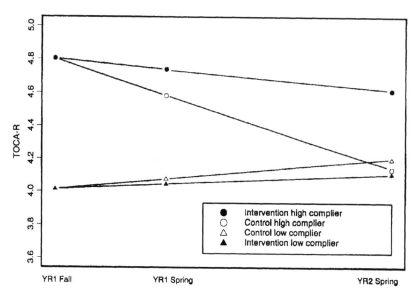

FIG. 6.3. Estimated mean curves of readiness to learn: CACE analysis using the growth model approach.

the cutpoint that is essential for CACE analysis, given that there are no established methods that can determine an optimal cutpoint. Without information from exploratory growth mixture analysis, one may simply choose to categorize 50% of individuals into the high complier category and 50% of individuals into the low complier category. In the JHU PIRC preventive trial example, the CACE analysis based on the cutpoint from exploratory growth mixture analysis was found to be substantially better than the CACE analysis based on the simple categorization (i.e., 50:50) in terms of model fit, precision of intervention effect estimates, and precision in classification of individuals into different compliance categories.

CONCLUSION

This study demonstrated the estimation of differential average intervention effects at varying levels of compliance in a growth mixture modeling framework, where the effect of the intervention is defined based on a trend or a growth trajectory of individuals. It was demonstrated in simulation studies that the quality of intervention effect estimates in ANCOVA and growth model approaches is very close, implying comparability of the two approaches. The growth model approach, however, showed slightly better point estimates and standard errors than did the ANCOVA approach.

Although the gap between the two approaches in point estimates and standard errors was not dramatic, it still resulted in a noticeable difference between the two methods in terms of statistical power to detect intervention effects. In the growth model approach, precision of average causal effect estimates can be improved by handling measurement errors and by utilizing trajectory information. In contrast, the ANCOVA approach only considers the outcome measured at the first and the last time point, and ignores the outcome measured in between. The loss in information may lead to lower precision in the ANCOVA approach.

In the JHU PIRC preventive trial example shown in this study, the differential effect of intervention was estimated through CACE analysis using ANCOVA and growth model approaches. CACE analyses using ANCOVA and growth mixture approaches showed a larger effect of intervention for high compliers compared to the overall effect in the ITT method. The results were also compared between CACE analyses using the ANCOVA and growth model approaches. In line with simulation study results, it was shown in this example that ANCOVA and growth model approaches have close intervention effect estimates, but CACE analysis using the growth model approach showed a slightly tighter confidence interval than CACE analysis using the ANCOVA approach. It was also demonstrated that the parameterization of the growth model approach adds more flexibility in modeling and provides richer information than that of the ANCOVA approach.

In the JHU PIRC preventive trial example shown in this study, individuals were classified into two groups, and CACE models were identified based on various covariates and growth trajectories. The exclusion restriction could not be assumed in this example, because low compliers were also exposed to the intervention. The intervention might have had a weaker impact on low compliers, but it cannot be guaranteed that the intervention had no effect at all. Without assuming the exclusion restriction, the identifiability and the quality of CACE estimation relies on auxiliary information (Hirano et al., 2000; Jo, in press-a). Given that, it is desirable to use multiple sources of information to improve accuracy and efficiency in the estimation. In the JHU PIRC preventive trial example shown in this study, not only covariate information but also trajectory information was used to identify class membership and to increase efficiency in the estimation of differential intervention effects. Although previous research showed that it is possible to identify CACE models without assuming the exclusion restriction based on auxiliary information, very little is known about how this method should be applied in practice. More research is needed in this area to explore what kind of information and modeling approaches are more efficient and how stability of models should be checked in extended versions of CACE models.

This study also examined the possibility of using exploratory growth mixture analysis as a data-mining tool that precedes CACE analyses. The confirmatory mixture analysis (i.e., CACE analysis) following the exploratory mixture analysis is based on the idea that subpopulations that are already different in the absence of intervention will be more likely to differ in compliance behavior. How closely subpopulations derived by exploratory growth mixture analyses are related to subpopulations derived by CACE analysis varies in different situations. When the intervention condition includes many sessions, as in the JHU PIRC preventive trial, how individuals are categorized into different compliance classes is critical for CACE analysis. However, little is known about how to determine optimal cutpoints and number of cutpoints. Given that, exploratory growth mixture analysis can be useful at the planning stage of CACE analysis in the sense that it provides information about subpopulations that are heterogeneous in the absence of the intervention. Further research is needed in this area to establish a systematic way of connecting exploratory and confirmatory mixture analyses.

ACKNOWLEDGMENTS

This study was supported by National Institute on Alcohol Abuse and Alcoholism Grant K02 AA 00230 (Bengt Muthén, Principal Investigator) and National Institute of Mental Health Grant P50 MH38725 (Philip Leaf, Principal Investigator). We would like to thank Nicholas Ialongo and Sheppard Kellam for their data and insightful advice.

REFERENCES

Angrist, J. D., Imbens, G. W., & Rubin, D. B. (1996). Identification of causal effects using instrumental variables. *Journal of the American Statistical Association, 91*, 444–455.

Bloom, H. S. (1984). Accounting for no-shows in experimental evaluation designs. *Evaluation Review, 8*, 225–246.

Dempster, A., Laird, N., & Rubin, D. B. (1977). Maximum likelihood from incomplete data via the EM algorithm. *Journal of the Royal Statistical Society, 39*, (Series B), 1–38.

Efron, B., & Feldman, D. (1991). Compliance as an explanatory variable in clinical trials. *Journal of the American Statistical Association, 86*, 9–17.

Hirano, K., Imbens, G. W., Rubin, D. B., & Zhou, X. H. (2000). Assessing the effect of an influenza vaccine in an encouragement design. *Biostatistics, 1*, 69–88.

Holland, P. W. (1986). Statistics and causal inference. *Journal of the American Statistical Association, 81*, 945–970.

Holland, P. W. (1988). Causal inference, path analysis, and recursive structural equation models. In C. Clogg (Ed), *Sociological methodology* (pp. 449–484). Washington, DC: American Sociological Association.

Ialongo, N. S., Werthamer, L., Kellam, S. G., Brown, C. H., Wang, S., & Lin, Y. (1999). Proximal impact of two first-grade preventive interventions on the early risk behaviors for later substance abuse, depression and antisocial behavior. *American Journal of Community Psychology, 27*, 599–642.

Imbens, G. W., & Angrist. J. (1994). Identification and estimation of local average treatment effects. *Econometrica, 62*, 467–476.

Imbens, G. W., & Rubin, D. B. (1997). Bayesian inference for causal effects in randomized experiments with non-compliance. *The Annals of Statistics, 25*, 305–327.

Jo, B. (in press-a). Estimating intervention effects with noncompliance: Alternative model specifications. *Journal of Educational and Behavioral Statistics.*

Jo, B. (in press-b). Model misspecification sensitivity analysis in estimating causal effects of interventions with noncompliance. *Statistics in Medicine.*

Jo, B. (2002). Statistical power in estimating causal effects of interventions with noncompliance. *Psychological Methods, 7*, 178–193.

Jo, B., & Muthén, B. O. (2001). Modeling of intervention effects with noncompliance: A latent variable modeling approach for randomized trials. In G. A. Marcoulides & R. E. Schumacker (Eds.), *New developments and techniques in structural equation modeling* (pp. 57–87), Lawrence Erlbaum Associates.

Kellam, S. G., & Rebok, G. W. (1992). Building developmental and etiological theory through epidemiologically based preventive intervention trials. In J. McCord & R. E. Tremblay (Eds.), *Preventing anti-social behavior: Interventions from birth through adolescence* (pp. 162–195). New York: Guilford.

Little, R. J. A., & Rubin, D. B. (1987). *Statistical analysis with missing data.* New York: Wiley.

Little, R. J. A., & Yau, L. (1998). Statistical techniques for analyzing data from prevention trials: Treatment of no-shows using Rubin's causal model. *Psychological Methods, 3*, 147–159.

McLachlan, G. J., & Krishnan, T. (1997). *The EM algorithm and extensions.* New York: Wiley Series.

McLachlan, G. J., & Peel, D. (2000). *Finite mixture models*. New York: Wiley Serires.

Muthén, B. O. (2001a). Latent variable mixture modeling. In G. A. Marcoulides & R. E. Schumacker (Eds.), *New developments and techniques in structural equation modeling* (pp. 1–33), Lawrence Erlbaum Associates.

Muthén, B. O. (2001b). Second-generation structural equation modeling with a combination of categorical and continuous latent variables: new opportunities for latent class/latent growth modeling. In A. Sayer & L. Collins (Eds.), *New methods for the analysis of change* (pp. 291–322). Washington, DC: American Psychological Association.

Muthén, B. O., Brown, C. H., Khoo, S. T., Yang, C. C., & Jo, B. (1997). *General growth mixture modeling of latent trajectory classes: Perspectives and prospects*. Manuscript in preparation.

Muthén, B. O., Brown, C. H., Masyn, K., Jo, B., Khoo, S. T., Yang, C. C., Wang, C. P., Kellam, S. G., Carlin, J. B., & Liao, J. (in press). General growth mixture modeling for randomized preventive interventions. *Biostatistics*.

Muthén, L. K. & Muthén, B. O. (1998-2001). M*plus user's guide*. Los Angeles: Muthén & Muthén.

Muthén, B. O., & Shedden, K. (1999). Finite mixture modeling with mixture outcomes using the EM algorithm. *Biometrics, 55,* 463–469.

Nagin, D. S. (1999). Analyzing developmental trajectories: A semiparametric, group-based approach. *Psychological Methods, 4,* 139–157.

Rubin, D. B. (1974). Estimating causal effects of treatments in randomized and nonrandomized studies. *Journal of Educational Psychology, 66,* 688–701.

Rubin, D. B. (1978). Bayesian inference for causal effects: The role of randomization. *Annals of Statistics, 6,* 34–58.

Rubin, D. B. (1980). Discussion of "Randomization analysis of experimental data in the Fisher randomization test" by D. Basu. *Journal of the American Statistical Association, 75,* 591–593.

Rubin, D. B. (1990). Comment on "Neyman (1923) and causal inference in experiments and observational studies." *Statistical Science, 5,* 472–480.

Tanner, M. (1996). *Tools for statistical inference: Methods for the exploration of posterior distributions and likelihood functions*. New York: Springer.

Titterington, D. M., Smith, A. F. M., & Makov, U. E. (1985). *Statistical analysis of finite mixture distributions*. Chichester, England: Wiley.

Werthamer-Larsson, L., Kellam, S. G., & Wheeler, L. (1991). Effect of first-grade classroom environment on child shy behavior, aggressive behavior, and concentration problems. *American Journal of Community Psychology, 19*, 585–602.

West, S. G., & Sagarin, B. J. (2000). Participant selection and loss in randomized experiments. In L. Bickman (Ed.), *Research design* (pp. 117–154). Thousand Oaks, CA: Sage.

7

Analysis of Repeated Measures Data

Elizabeth R. Baumler
Ronald B. Harrist
*University of Texas-Houston
Health Science Center*

Scott Carvajal
*Mexican American Studies
and Research Center,
The University of Arizona*

Multilevel models are useful in analyzing repeated measures data from behavioral research studies. Such studies often follow a cohort of individuals over time, resulting in repeated measurements on individuals. A cohort design may result in increased statistical power to detect effects of interest in a given set of participants (Maxwell, 1998) and allows examination of change in the study outcome over time, thus providing more complete modeling of human behavior (Bryk & Raudenbush, 1992).

Repeated measurements within an individual are generally correlated. This correlation is sometimes called intraclass correlation (ICC). Such repeated measures data can be thought of as a special case of "nesting" of observations within individuals, resulting in a positive ICC. Multilevel models may be used to analyze data in the presence of intraclass correlation. Although statistical procedures for the analysis of repeated measures data such as ANOVA have been in existence for some time, most such procedures require regular, balanced measurement occasions. A multilevel model does not require this restrictive pattern and provides statistically efficient parameter estimation for any pattern of measurements (Goldstein, 1995). This is particularly useful in the analysis of data resulting from behavioral interventions.

It is also common in follow-up studies for the repeated measurements to be obtained at irregular intervals, either by design or due to missing of planned data collection. Data are particularly likely to be missing when the period of data collection extends over several years. Multilevel models are flexible enough to deal with such missingness of data without resorting to a listwise deletion or casewise deletion of records as may be required by standard analysis of variance techniques such as repeated measures ANOVA or ANCOVA (Bryk & Raudenbush, 1992; Goldstein, 1995). Deletion of data to achieve patterns that can be analyzed by standard techniques can lead to substantial loss of information and can introduce bias into the analysis. Supplementing data by imputation or other data modeling techniques to achieve desired patterns may also result in a biased analysis. Even the most computationally sophisticated and elegant of such techniques requires additional, and usually unverifiable, assumptions regarding the nature of the missing observations (Graham & Schafer, in press).

Multilevel models can be effectively used almost without regard to the patterns of missingness, provided data are missing at random, that is, that missingness is not related to the variables being measured. Caution in interpretation is advised if missingness is associated with the magnitude of the outcome variable or a predictor variable, or if data are missing due to other nonrandom causes.

MULTILEVEL REPEATED MEASURES MODEL

Consider a data set with n repeated observations taken on m students. Let $i = 1, 2, \ldots n$ denote the occasion of observation and $j = 1, 2, \ldots m$ denote the student. The n repeated measurements taken on the jth student are likely to be correlated, thus violating the assumption of independence between observations. This type of repeated measures data can be modeled using a multilevel model in which the level of nesting occurs within the individual.

Let y_{ij} denote the ith observation on the jth student for a normally distributed dependent variable. The variance in y_{ij} can be thought of as being partitioned into two parts, variation that occurs between repeated measurements but within a student and variation that occurs between students. The following simple two–level multilevel model with a random intercept only expresses the dependent variable y_{ij} as a function of a single linear predictor

$$y_{ij} = \beta_{oj} + \beta_1 x_{ij} + (u_j + e_{ij}) \qquad (7.1)$$

where u_j denotes the random error associated with the student level variation, and e_{ij} denotes the random error associated with the variation between repeated measurements for each student. In this model, the slope is not allowed to be random. Here, β_{0j} and β_1 are unknown parameters denoting the mean and intercept, respectively, of the simple linear two-level model. Each of the error terms is assumed to be identically and independently normally distributed with mean 0 and variances equal to σ_u^2 and σ_e^2 at the student and occasion levels, respectively. It should be noted that methods exist for modeling more complex error distributions (see Goldstein, 1995), however, such models are beyond the scope of this chapter. In this model, the dependent variable y_{ij} is expressed as a linear function of an intercept β_{oj} and a single independent variable x_{ij} that varies both by student and by repeated measurement. In this model, only the intercept is allowed to be random at the student level. This means that the trajectory of the repeated measurements for each student is assumed to have the same slope, β_1, but the intercept, β_{0j}, is allowed to vary from student to student. Inclusion of a random intercept into the model makes it appropriate for modeling correlated data such as repeated measurements. By including a random intercept, both the variance between students as well as the variance between repeated observations is estimated. If the intercept were to be fixed, not allowed to be random, the multilevel model would be reduced to a simple least squares regression model in which all observations are assumed to be independent. Statistical models that assume independence of observations fail to take into account the correlations induced by grouping, or in this case repeated observations, often resulting in an underestimation of the standard error of the parameter estimates and provide inefficient estimates of the parameters of interest (Goldstein, 1995; Murray, 1998). Such inappropriate statistical inferences could have profound influence on scientific conclusions and on social and behavioral science policy (Rooney & Murray, 1996).

Independent (predictor) variables, which vary only from student to student, remaining constant between measurement occasions, may also be entered into the model. For example, gender and ethnicity remain constant between measurement occasions but vary from student to student.

In repeated measures studies, time is usually of particular importance. The model can be easily extended to include a measure of time as well as other relevant covariates. Let t_{ij} denote the time, on some appropriate scale, of the ith measurement on the jth student. Let $x_{1j}, \ldots, x_{pj}$, $p = 1, 2, \ldots s$ where s is the number of independent variables, such as sex or race, that vary at the student level. The dependent variable can then be expressed as a function of time as well as of the additional covariates in the following way.

$$y_{ij} = \beta_{oj} + \beta_1 t_{ij} + \beta_2 x_{1j} + \beta_3 x_{2j} + \cdots \beta_p x_{pj} + (u_j + e_{ij}) \qquad (7.2)$$

The regression coefficients $\beta_2, \ldots \beta_{p+1}$ measure the average linear relationships between the independent variables x_{1j} , $\ldots x_{pj}$ and the dependent variable. Here, the regression parameter β_1 measures the change in the dependent variable y_{ij} per unit change in time, t_{ij}.

In Equation 7.2 time is expressed as a continuous measure. The model can also be set up to incorporate a discrete measure of time. Let t_{ij} take on four discrete values (0, 1, 2, 3) denoting the baseline, first, second, and third follow-up occasions. Using the baseline as the referent category, indicator variables can be created in the following way to reflect the four measurement occasions.

$$
\begin{array}{llll}
t_{1ij} = 1 & if & t_{ij} = 1; & 0 \quad otherwise \\
t_{2ij} = 1 & if & t_{ij} = 2; & 0 \quad otherwise \\
t_{3ij} = 1 & if & t_{ij} = 3; & 0 \quad otherwise.
\end{array}
$$

The model in Equation 7.2 can then be rewritten to incorporate the discrete time measurement as follows

$$
\begin{aligned}
y_{ij} = & \ \beta_{oj} + \beta_1 t_{1ij} + \beta_2 t_{2ij} + \beta_3 t_{3ij} \\
+ & \ \beta_4 x_{1j} + \beta_5 x_{2j} + \cdots \beta_{p+3} x_{pj} + (u_j + e_{ij}).
\end{aligned}
\tag{7.3}
$$

Here the estimated coefficients β_1, β_2, and β_3 measure the relationship between the follow-up occasion and the dependent variable relative to the first measurement. For example, suppose we are modeling four repeated observations of percent body fat taken on students in a fitness program. A negative value for β_1 would indicate that the average percent body fat of the students was lower at second follow-up observation than at the first. This type of modeling, which incorporates a discrete measure of time, closely parallels the repeated measures ANOVA model.

Such multilevel modeling provides a flexible method for analysis of repeated measures data in that no assumptions are made regarding the timing between observations nor the number of observations per individual. Further, multilevel models can be extended to encompass more complex data structures, such as repeated measurements taken on students nested within schools (Goldstein, 1995).

Multilevel repeated measures models can also be fitted when the dependent variable has a distribution other than normal, such as binomial or poisson. There are several estimation algorithms in common use for fitting such nonlinear random effects models. One such estimation method, known as penalized quasi-likelihood (PQL), provides relatively unbiased parameter estimates for many of the multilevel models commonly used (Goldstein & Rasbash, 1996). It should be noted that PQL estimation methods are computationally intensive and sometimes fail to converge (Goldstein, 1995). In some cases, marginal quasi-likelihood (MQL) estimation can be used, which is less computationally intensive. However, caution should be exercised when using MQL as, under some conditions, it produces biased parameter estimates (Rodrigues & Goldman, 1995). In practice, often MQL can be used initially to generate reasonable starting values and the model will then converge using PQL. Software for fitting multilevel models has become increasingly more available. Some of the software packages currently available include, MLwiN (Multilevel Models Project - Woodhouse, Rasbash, Goldstein, Yang & Plewis, 1996b), SAS PROC MIXED (SAS Institute, Inc., 2002), MIXREG and MIXOR (Hedeker and Gibbons, 1996), VACRL (Longford,1988), BUGS (Spiegelhalter, Thomas, Best, & Gilks, 1995), MLA (Busing & Van der Leeden, 1994) and HLM (Bryk, Raudenbush, Seltzer, & Congdon, 1988).

EXAMPLE 1

The Safer Choices study was an evaluation of a theory-based multicomponent HIV/STD and pregnancy prevention program designed to reduce sexual risk–taking behavior among high school students. The Safer Choices intervention was designed to reduce such behaviors by changing theory-derived psychosocial determinants of sexual behavior (Bandura, 1986; Bartholomew, Parcel, & Kok, 1998; Basen-Engquist et al., 1999; Basen-Engquist & Parcel, 1992; Fishbein, Middlestadt, & Hitchcock, 1994). The evaluation of this project consisted of both a cohort assessed over 4 years and three successive cross-sectional samples on the key hypothesized behaviors of influence and psychosocial mediators of behavior change.

The Safer Choices intervention was implemented during the 1993 to 1994 and 1994 to 1995 school years. The evaluation utilized a randomized trial involving 20 schools—10 schools in southeast Texas and 10 in northern California. Within each site, 5 of the schools were randomly assigned to the multiple component program (Safer Choices); the remaining half were assigned to the comparison program (a standard knowledge-based curriculum). To assess the effectiveness of the intervention, cohort data were collected at all 20 schools on a cohort of 3,869 students who were followed for 2 years. The data included one baseline measure taken prior to the intervention, and follow-up measures taken at 7, 19, and 31 months postintervention. Because randomization occurred at the school level and the repeated observation was the unit of analysis, data were clustered both within schools and within students.

Multilevel modeling procedures were used to assess the average intervention effect in the cohort sample across time. Three-level models were used, where level-1 was the repeated measurement, level-2 the student, and level-3 the school. This type of modeling allowed estimation of the intervention effect over time and also allowed retention of students with incomplete follow-up information in the cohort.

The model described in this example was used to evaluate the impact of Safer Choices intervention on the condom self-efficacy outcome. The measure of condom self-efficacy was defined to be the average of three Likert-type items ("totally sure" to "not sure at all") expressing the respondents' confidence that they could use a condom correctly and explain to their partner how to use a condom correctly (Chronbach's $\alpha = .61$). By averaging the items, the scale took on a continuous range of values from 1, being the lowest score, to 3, being the highest. This variable was approximately normally distributed in this sample. Additional psychometric and validation findings based on confirmatory factor analyses concerning this and all other psychosocial scales are available in Basen-Engquist Parcel, and Kok, 1998; Basen-Engquist et al. (1999).

The model used to evaluate the intervention's impact on this outcome measure included the participants' baseline responses on the outcome to adjust for imbalances between the intervention and control condition at baseline, group assignment (intervention or control), location (Texas or California), measurement occasion (first, second, or third follow-up), and a set of outcome-specific covariates. Outcome-specific covariates were included in the initial model if they were judged a priori to be both plausibly related to the outcome under consideration, and were unevenly distributed among the intervention and control conditions. They were retained in the final model if they remained statistically significant in the final stage of multilevel modeling.

First, consider a simple multilevel model with the repeated observation as the unit of analysis. For simplicity, we first consider only the independent variables measuring the baseline measure of the outcome, follow-up observation number, and treatment condition. Because primary interest was in the impact of the intervention, the baseline measurement was used for the purposes of adjusting for differences in the dependent measure that existed prior to implementation of the intervention. Therefore, it was included in the model as an independent variable. The baseline measure was not centered. A three–level model that allows the intercept for each school to be expressed as the sum of the average intercept (over all students and all schools) and two random deviations, one that varies within students and one that varies only between schools, can be expressed as follows. Let y_{ijk} denote the condom self-efficacy score at the ith observation on the jth student in the kth school. For observations within the ith student, the model is expressed as

$$y_{ijk} = \beta_{0jk} + \beta_1 x_{1ijk} + \beta_2 x_{2ijk} + \beta_3 x_{3ijk} + \beta_4 x_{4k} + \varepsilon_{ijk}, \qquad (7.4)$$

where:

x_{1ijk} = baseline score for the jth student
x_{2ijk} = 1 if it is the 2nd follow-up for j^{th} student; 0 otherwise
x_{3ijk} = 1 if it is the 3rd follow-up for the jth student; 0 otherwise
x_{4k} = 1 if the school is in the treatment group; 0 otherwise

and

β_{0jk} = mean score for first follow-up for the ij^{th} student
β_1 = effect of baseline score
β_2 = effect of 2nd follow-up
β_3 = effect of 3rd follow-up
β_4 = intervention effect (mean across time and over all schools)

The predictors denoted by the x's are indicators of student or school characteristics and the β's are unknown regression coefficients to be estimated from the data. It is assumed, subject to verification, that the random variables ε_{ijk} are independently distributed as $N\left(0, \sigma_\varepsilon^2\right)$. This part of the model is simply a multiple linear regression model conditioned on being restricted to observations within the jk^{th} student. Now, the intercept can be allowed to vary over students by writing

$$\beta_{ojk} = \beta_{ok} + u_{jk}, \qquad (7.5)$$

where β_{oj} denotes the school mean of all the student means for the kth school, β_{ojk}, and u_{jk} is the random deviation of the jth student's mean within the kth school. By assumption, again subject to verification, the level–2 random variable u_{jk} follows the distribution $N\left(0, \sigma_u^2\right)$ independently of the ε_{ijk}'s of level–1. Similarly, the intercept can be allowed to vary over schools by expressing the intercept β_{ojk} as a function of the mean of all the school means, β_o, and a random deviation of the kth school's mean, v_k. This is expressed in the following way:

$$\beta_{ok} = \beta_o + v_k. \qquad (7.6)$$

By assumption, again subject to verification, the level–3 random variable v_k is assumed to follow the distribution $N(0, \sigma_v^2)$ independently of the u_{jk} and ε_{ijk}'s of level–2 and level–1. These level–1, level–2, and level–3 components may be combined to form the complete three-level model:

$$y_{ijk} = \beta_{0jk} + \beta_1 x_{1ijk} + \beta_2 x_{2ijk} + \beta_3 x_{3ijk} + \beta_4 x_{4k} + (v_k + u_{jk} + \varepsilon_{ijk}), \qquad (7.7)$$

TABLE 7.1

Impact of Intervention on Condom Self–Efficacy

	Estimate (SE)
Fixed	
Intercept (β_{ojk})	1.29 (0.03)
Baseline Score (β_1)	0.44 (0.01)
2nd Follow-up (β_2)	0.07 (0.009)
3rd Follow-up (β_3)	0.14 (0.009)
Group Assignment (β_4)	0.16 (0.02)
Random	
σ_v^2(variation between schools)	0.001 (0.001)
σ_u^2 (variation between students)	0.085 (0.003)
$\sigma\varepsilon^2$ (variation between observations)	0.129 (0.002)

subject to the distributional assumptions mentioned earlier. Fitting the model involves estimating the unknown β's and the unknown variances σ_v^2, σ_u^2 and σ_ε^2. The intervention's overall effectiveness across time, students, and schools can be determined by testing the statistical hypothesis that β_4 equals zero. This model can easily be extended by designating the group coefficient to be random at level–3. This allows the impact of the intervention to differ from school to school (Raudenbush, 1997). Although this model could provide useful information regarding the population under consideration, it was not appropriate in the overall evaluation of the intervention effect in this study. This modeling approach made it possible to test the impact of the intervention with test statistics corrected for the clustering of observations within students and students within schools. The parameter estimates resulting from estimation of the model specified in Equation 7.7 are presented in Table 7.1. A complete analysis of the cohort data, including a multilevel analysis of all psychosocial, behavioral, and demographic variables are presented in Coyle et al. (1999) and Coyle et al. (in press).

Of primary interest in Table 7.1 is the estimate of β_4, 0.16, and its estimated standard error, 0.02. Recall from the model that β_4 is the coefficient to the indicator variable of group assignment. The group assignment variable was coded 1 for intervention and 0 for control. Under this parameterization, β_4 can be interpreted as the average intervention effect across the three study observations after adjusting for other covariates in the model. Condom self-efficacy was scored on a scale of 1 to 3. Therefore, an effect of 0.16 indicates that the average adjusted mean self-efficacy score across time in the intervention group was 0.16 higher in the intervention group than in the control group. Under the distributional assumptions mentioned earlier, the ratio of the estimate of β_4 to its standard error approximately follows a Student's t-distribution with degrees of freedom approximately equal to the number of observations minus the number of parameters estimated in the model. This permits us to test the hypothesis that $\beta_4 = 0$, or, equivalently, to determine the associated p value. Such tests of the model's coefficients are called *Wald tests*, after their originator, and make use of the properties of maximum likelihood estimation. Even when the underlying assumptions of normality are violated, such tests may still be useful, because the ratio will follow the normal distribution approximately, provided there is a fairly large number of observations. Because the sample size for this study is very large, the degrees of freedom for the t statistic is vary large, and its distribution under the null hypothesis is approximately $N(0,1)$. The ratio is 8.00 and the p value as determined by reference to the standard normal distribution is less than 0.001. This leads to the conclusion that the intervention had a statistically significant effect on the average condom self-efficacy score across the three follow-up observations. Similar inspection of the other parameter estimates in Table 7.1 shows that the effect of baseline score, and indicator variables for the second follow-up, and third follow-up are all significantly related to the dependant variable as well.

The intraclass correlation (ICC) is the proportion of the total variance that occurs between observations within a level of clustering. The ICC that expresses the strength of the positive correlation between the responses of students within the same school (level 3) is thus (Raudenbush, 1997):

$$ICC_3 = \frac{\sigma_v^{\ 2}}{\sigma_v^{\ 2} + \sigma_u^{\ 2} + \sigma_\varepsilon^{\ 2}}. \qquad (7.8)$$

For condom use efficacy, the ICC_3 for students within the same school is estimated as $ICC_3 = 0.001 / (0.001 + 0.085 + 0.129) = 0.005$, suggesting a small, but still possibly important, influence of the clustering. The ICC that expresses the strength of the positive correlation between the repeated observations within the same student is defined as

$$ICC_2 = \frac{\sigma_u^{\ 2}}{\sigma_v^{\ 2} + \sigma_u^{\ 2} + \sigma_\varepsilon^{\ 2}}. \qquad (7.9)$$

For the same outcome, the ICC for repeated observations within the same student is estimated as $ICC_2 = 0.085 / (0.001 + 0.085 + 0.128) = 0.40$. As would be expected, there is a high degree of intraclass correlation among repeated observations within the same student. It is often found that in cases where the intraclass correlation between repeated measures is high,

TABLE 7.2

Repeated Measures Data Structure FOR ANOVA

Student	Baseline	GPA 1st Follow-up	GPA 2nd Follow-up	GPA 3rd Follow-up	GPA 4th Follow-up
1	3.2	3.5	3.8	3.2	3.3
2	2.9	2.5	2.8	3.0	*
3	4.0	3.9	*	3.8	3.8
4	3.1	3.3	3.6	3.5	3.6
5	2.3	*	2.6	*	2.4
6	3.5	3.6	3.2	*	3.3
7	3.8	3.7	3.9	3.8	3.6
8	2.7	2.6	2.5	2.6	2.5

the majority of the variation present in the sample occurs at a higher level, such as the student or school level (Goldstein, 1995). The model expressed in Equation 7.7 can easily be extended to include covariates measured at the school, student, or observation level. Additionally, group assignment by time interaction terms can be added to explore whether the intervention had a differential impact at each follow-up observation.

MISSING OBSERVATIONS

One advantage of using a multilevel model for analysis of repeated measures data is the ability to use incomplete cases or mistimed follow-up measurements in the analysis. Although statistical procedures such as ANOVA may be used to analyze some repeated measures data, the multilevel models analysis uses maximum likelihood rather than least squares methods for estimating the unknown coefficients of the model and thus may be used even for unbalanced designs or for irregularly spaced observations (Goldstein, 1995). This feature is particularly useful in the analysis of data resulting from behavioral interventions.

Consider the repeated measures data presented in Table 7.2 that represent the self-reported grade point average of 8 students observed at baseline and on four subsequent occasions at 6–month intervals.

As often occurs, complete information is not available on all students for all observation times. In a repeated measures ANOVA model, a missing data point requires that either the data from the individual's entire record be omitted or the missing information imputed. Deletion of entire records from the analysis can result in a sizable loss of information. For the data represented in Table 7.2, only 4 of the 8 students have complete

TABLE 7.3
Repeated Measures Data Structure MLM

Student	Follow-up	GPA	Baseline
1	1	3.5	3.2
1	2	3.8	3.2
1	3	3.2	3.2
1	4	3.3	3.2
2	1	2.5	2.9
2	2	2.8	2.9
2	3	3.0	2.9
2	4	*	2.9
3	1	3.9	4.0
3	2	*	4.0
3	3	3.8	4.0
3	4	3.8	4.0
4	1	3.3	3.1
4	2	3.6	3.1
4	3	3.5	3.1
4	4	3.6	3.1
:	:	:	:
:	:	:	:
8	1	2.6	2.7
8	2	2.5	2.7
8	3	2.6	2.7
8	4	2.5	2.7

information. Deletion of the other students in the sample could result in possibly biased results. Although several methods of imputation exist to deal with missing data points, most require assumptions that are difficult to verify and produce distributional effects on the estimates that may be difficult to ascertain.

In an MLM approach to analyzing such repeated measures data, we are not limited to using data only from subjects with complete information. In a multilevel data structure, listwise deletion of missing values occurs at the observation level rather than at the individual record level. The multilevel analysis utilizes the data in the format expressed in Table 7.3, rather than the structure shown in Table 7.2.

In this format, all of an individual's available information is retained in the analysis. For example, student #2 would contribute to the estimation

TABLE 7.4

Safer Choices Attrition Rate

	Sample Size	% of Initial Cohort
Baseline cohort	3,869	-
1st Follow-up	3,677	95%
2nd Follow-up	3,212	83%
3rd Follow-up	3,058	79%
Complete follow-up information	2,788	72%
2 follow-up observations	561	14%
1 follow-up observation	447	12%
No follow-up information	73	2%

of the average GPA in the sample at the first, second, and third follow-up, but would not contribute any information to the estimation of the average GPA at the fourth follow-up.

Although this approach to missing observations is useful and helps to minimize bias resulting from lost observations, it is only valid under the assumption that the observations are missing at random. Under this assumption, the power of the statistical tests may be improved because observations are utilized that may have otherwise been discarded. It should be noted, however, that a multilevel analysis does not offer a solution to the bias and lack of power that can occur when a great many observations are missing from the sample nor is it a substitute for proper study design and effective retention of study subjects.

EXAMPLE 2

Consider the Safer Choices study from Example 1. Table 7.4 summarizes the attrition patterns and rates for the study from the baseline measure to the 31-month follow-up.

A total of 3,677 of the 3,869 students (95%) in the cohort were surveyed at first follow-up. Although extensive follow-up protocols were in place, the retention rate dropped to 83% (3,212) at the 19–month followup and to 79% (3,058) by the final 31-month observation. Of the 3,869 students randomized into the study, complete information was obtained on 2,788 (72%) of the subjects. An additional 561 (14%) of the sample had two of the three follow-up observations with 447 (12%) providing information from only one of the followup periods. If the analysis were to be conducted on only those individuals with complete information, 1,081 (28%) of the students in the cohort would be eliminated entirely. By using an MLM approach, 1,008 (26%) students with partial information were utilized in the analysis contributing information to the estimation of the regression parameters at the time points for which they provided data.

To illustrate how the use of partial information can impact estimation of the regression parameters, consider the outcome measure of condom self–efficacy as discussed in Example 1. The results presented in Table 7.1 reflect the inclusion of students with incomplete information into the analysis. More specifically, this analysis reflects the inclusion of 3,532 observations at the 7–month follow-up; 3,097 observations at the 19–month follow-up, and 2,978 observations at the 31–month follow-up. Next, consider the same analysis conducted on only those students with observations at all three follow-up periods. Excluding students with incomplete data results in a sample size of 2,664 students and 7,992 observations. The model specified in Equation 7.7 was fit to the data set after a listwise deletion of all students who did not have follow-up information at all three time points. A comparison of the results from the original analysis that included students with incomplete information and one that did not can be found in Table 7.5.

Excluding cases with incomplete follow-up information resulted in slight changes in the parameter estimates and their standard errors. For the coefficients measuring the relationship between the dependant variable and the timing of the follow-up (Time2, Time3) the estimates remained the same ($\beta_2 = 0.07$, $\beta_3 = 0.14$), however, the standard error for both increased slightly from 0.009 in the analysis that included students with incomplete follow-up to 0.01 in the analysis that was conducted only on students with complete data. The standard errors remained the same for all other fixed parameters in the model, however the estimates of these parameters did change somewhat. The effect of the baseline score, β_1, increased slightly from 0.45 to 0.44 whereas the estimated average intercept for the model, β_o, increased from 1.29 to 1.25. Of the most importance perhaps is the observed change in the estimated effect of the treatment group assignment, β_4, which increased from 0.16 to 0.19, whereas the estimate of the standard error remained the same in both analyses. This change in estimated treatment effect did not result in a change in the test; however, had the effect been marginal, the change in the regression coefficient could have led to a false positive result, that is, rejection of the null hypothesis when it was, in fact, true.

TABLE 7.5
Comparison of Including and Excluding Incomplete Data

	Partial Data Included	Complete Data Only
Total number of students	3,719	2,664
Number of 1st follow-up observations	3,532	2,664
Number of 2nd follow-up observations	3,097	2,664
Number of 3rd follow-up observations	2,978	2,664
MODEL	Estimate (SE)	Estimate (SE)
Fixed:		
Intercept (β_{ojk})	1.29 (0.03)	1.25 (0.03)
Baseline score (β_1)	0.44 (0.01)	0.45 (0.01)
2nd follow-up (β_2)	0.07 (0.009)	0.07 (0.01)
3rd follow-up (β_3)	0.14 (0.009)	0.14 (0.01)
Group assignment (β_4)	0.16 (0.02)	0.19 (0.02)
Random:		
σ_v^2 (variation between schools)	0.001 (0.001)	0.001 (0.001)
σ_u^2 (variation between students)	0.085 (0.003)	0.086 (0.004)
$\sigma\varepsilon^2$ (variation between observations)	0.129 (0.002)	0.126 (0.002)

SUMMARY

Application of multilevel models to the analysis of repeated measures cohort data provides flexibility both in their ability to model a wide range of distributions and study designs and for dealing with missing data. Cohort studies that follow individuals over an extended period of time, such as the Safer Choices study described in this chapter, are often plagued with missing or incomplete follow-up information on study subjects. A multilevel modeling approach provides an alternative to such analytic techniques as a repeated measures analysis of variance, which require a listwise deletion of individuals with incomplete information. Data from the Safer Choices study were used to illustrate how multilevel modeling procedures can be used to assess the across-time intervention effect in a repeated measures cohort study design. This modeling technique utilized data from students in the cohort with incomplete follow-up information, thus allowing all students to be included in the analysis.

It should also be noted that multilevel models are related to many other modeling approaches typically employed in behavioral and psychological research, such as structural equation modeling (Kaplan & Elliot, 1997; Muthén, 1994), proportional hazards regression (Carvajal et al., in press; Goldstein 1995), mediational analysis (Krull & MacKinnon, in press), and probit, multinomial, logistic, or Poisson models (Gibbons & Hedeker 1997; Goldstein 1995). However, MLMs may not be ideal for all study designs where nonindependence in the data may be encountered.

REFERENCES

Bandura, A. (1986). *The social foundations of thought and action: A social-cognitive theory.* Englwood Cliffs, NJ : Prentice–Hall.

Bartholomew, L. K., Parcel, G. S., & Kok, G. (1998). Intervention mapping: A process for developing theory and evidence-based health education programs. *Health Education and Behavior, 25*(5), 545–563.

Basen-Engquist, K., & Parcel, G. (1992). Attitudes, norms and self-efficacy: A model of adolescents' HIV-related sexual risk behavior. *Health Education Quarterly, 12*, 263–277.

Basen-Engquist, K., Mâsse, L., Coyle, K., Kirby, D., Parcel, G., Banspach, S., & Nodora, J. (1999). Validity of scales measuring the psychosocial determinants of HIV/STD related risk behavior in adolescents. *Health Education Research, 14*, 101–114.

Bryk, A., Raudenbush, S. W., Seltzer, M., Congdon, R. (1988). *An introduction to HLM: Computer program and user's guide.* Chicago: University of Chicago; Deparment of Education.

Bryk, A. S., & Raudenbush, S. W. (1992). *Hierarchical linear models: Applications and data analysis methods.* Newbury Park, CA: Sage.

Busing, F., & Van der Leeden, R. (1994). *MLA: Users guide* (Version 1.0b. University of Leiden, The Netherlands.

Carvajal, S. C., Parcel, G. S., Basen-Engquist, K. B., Banspach, S., Coyle, K., Kirby, D., & Chan, W. (in press). Predicting the onset of sexual intercourse in late adolescents with theory-based psychosocial determinants. *Health Psychology.*

Coyle, K., Basen-Engquist, K., Kirby, D., Parcel, G., Banspach, S., Collins, J., Baumler, E., Carvajal, S., Harrist, R. (in press). Safer choices: Long-term impact of a multi-component school-based HIV, other STD and pregnancy prevention program. *Public Health Reports.*

Coyle, K., Basen-Engquist, K., Kirby, D., Parcel, G., Banspach, S., Harrist, R., Baumler, E., Weil, M. (1999). Short term impact of a multi-component, school-based HIV/STD and pregnancy prevention program. *Journal of School Health, 69* (5), 181–188.

Fishbein, M., Middlestadt, S., & Hitchcock, P. (1994). Using information to change sexually transmitted disease-related behavior: An analysis based on the theory of reasoned action. In R. J. DiClemente & J. L. Peterson (Eds.), *Preventing AIDS: Theories and methods of behavioral interventions* (pp. 61–78). New York: Plenum Press.

Gibbons, R. D., & Hedeker, D. (1997). Random effects probit and logistic regression models for three level data. *Biometrics, 53*, 1527–1537.

Goldstein, H. (1995). *Multilevel statistical models.* New York: Wiley.

Goldstein, H., & Rasbash, J. (1996). Improved approximations for multilevel models with binary responses. *Journal of the Royal Statistical Society, 159*, 505–513.

Graham, J. W., & Schafer, J. L. (in press). On the performance of multiple imputation for multivariate data with small sample size. In: Hoyle, (Ed.), *Statistical strategies for small sample research.* Thousand Oaks, CA: Sage.

Hedeker, D., & Gibbons, R. D. (1996). MIXOR: A computer program for mixed-effects ordinal regression analysis. *Computer Methods and Programs in Biomedicine, 49*, 157–176.

Hedeker, D., Gibbons, R. D. (1996). MIXREG: A computer program for mixed-effects regression analysis with autocorrelated errors. *Computer Methods and Programs in Biomedicine, 49*, 229–252.

Kaplan, D., & Elliot, P. R. (1997). A model-based approach to validating education indicators using multilevel structural equation modeling. *Journal of Educational and Behavioural Statistics, 22*, 323–347.

Krull, J. L., & MacKinnon, D. (in press). Mediation in a multi-level (HLM) framework. *Evaluation Review.*

Longford, N. T., (1988). *VACRL - Software for variance component analysis of data with hierarchically nested random effects (maximum likelihood).* Princeton, NJ: Educational Testing Service.

Maxwell, S. E. (1998). Longitudinal designs in randomized group comparisons: When will intermediate observations increase statistical power? *Psychological Methods, 3*, 275–290.

Murray, D. M. (1998). *Design and analysis of group-randomized trials.* New York: Oxford University Press.

Muthén, B. O. (1994). Multilevel covariance structure analysis. *Sociological Methods and Research, 22*, 376–398.

Raudenbush, S. W. (1997). Statistical analysis and optimal design in cluster randomized trials. *Psychological Methods, 2*, 173–185.

Rodrigues, G., & Goldman, N. (1995). An assessment of estimation procedures for multilevel models with binary responses. *Journal of the Royal Statistical Society, 158,* 73–89.

Rooney, B. L., & Murray, D. M. (1996). A meta-analysis of smoking prevention programs after adjustment for errors in the unit of analysis. *Health Education Quarterly, 23,* 48–64.

Spiegelhalter, D. J., Thomas, A., Best, N. G., Gilks, W. R. (1995a). *BUGS: Bayesian inference using gibbs sampling* (Version 0.50). Biostatistics Unit, Cambridge, UK: MRC.

Woodhouse, G., Rasbash, J., Goldstein, H., Yang, M., Plewis, I., (1996). *Multilevel modeling applications: A guide for users of MLn.* University of London, London: Multilevel Models Project Institute of Education.

8

The Development of Social Resources in a University Setting: A Multilevel Analysis

Nicole Bachmann
University of Bern, Switzerland

Rainer Hornung
University of Zurich, Switzerland

Many students experience the beginning of their studies at university as a time that greatly challenges their adaptation skills and ability to cope with stress. Most freshmen have to manage in a completely new environment, stay for the first time in a place that is away from their parents, create new social contacts, and adapt to a new learning approach. A large number of students experience this transitory period as a kind of crisis that can be handled more or less successfully. In a longitudinal study about the effects of strain and resources on the health of students (N = 1,384) during the first year of studies, we have found that: (1) good social relationships among students studying within the same subject area constituted a central protective factor for their health, and that (2) the existence of social resources varied significantly depending on the area of study (Bachmann, Berta, Eggli, & Hornung, 1999). The social networks among students in the social sciences and arts were smaller, and they reported less satisfaction with their support in comparison to those studying natural and technical sciences.

The finding that social resources are an important contribution to the protection of health in stressful situations is not a new fact. The importance of social resources for health related behavior, the ability to cope with crises, the general health status, and even life expectancy have been established in a large number of studies (reviewed in Leppin & Schwarzer, 1990). What is particularly interesting, however, is the extent to which this important health resource developed depending on the subject studied.

With respect to this finding, two fundamental questions can be asked: (1) Is the observed phenomenon due to a selection effect, that is, that before their studies, the students already had different dispositions toward the creation of social contacts, such that the differences between areas of study were a consequence of selection at the moment of choosing a major (compositional effect); or (2) were the areas of study linked to different contextual conditions, thus having a more or less inductive effect on the creation of social contacts and reciprocal support among students (contextual effect).

The following question addresses the compositional effect: Do students of philosophy have so few contacts because they have a particularly low social competence? Whereas the following question addresses the contextual effect: Are the larger social networks found at technical studies due to the fact that, because of a comparatively higher number of required lectures, students have to spend more time together?

During the last decade, a new concept the transactional or ecological model of social resources (Hobfoll & Freedy, 1990; Hornung & Gutscher, 1994; Vaux, 1990), was developed within the area of theory and research about resources relevant to health. This new concept was based on a transactional process with reference to origin, maintenance, and use of social resources. According to this approach, social resources are defined as multidimensional with structural, functional, and evaluative components. The structural component describes the social network with its number and type of existing relationships, the functional component describes the expected or received social support, and the evaluative component represents the evaluation, that is, the degree of satisfaction with the social network and its support.

In a transactional approach, social support is defined as a constant dynamic process between an individual and his or her social environment. The creation, use, and maintenance of social resources requires an active interaction of the individual with his or her environment (Vaux, 1990). If we base our approach on this model, the differences observed between the groups of subjects studied should be due to both the compositional effect (i.e., depending on the area of study, the characteristics of the future students were already different before they started university studies), and the contextual effect (i.e., the characteristics of the areas of study are different and thus have an influence on the development of social relationships among the students).   Finally, the transactional theory includes the assumption that these two levels (individual and context) have a reciprocal effect on each other.   We could not test the assumption of reciprocity due to a lack of time series data.

According to the literature, the following characteristics of an individual are considered crucial for building and maintaining satisfactory and helpful social relationships; personal characteristics such as self-esteem, self-efficacy, empathy, network orientation (e.g., self-disclosure, family attachment) as well as gender and age (Barrera, & Baca, 1990; Burda, Vaux, & Schill, 1984; Hansson, Jones, & Carpenter, 1984; Lakey, McCabe, Fisicaro, & Drew, 1996; Tolsdorf, 1976).   On the contextual level, the following factors are discussed; cultural values, architectural aspect of the environment, spatial and temporal availability of personal contacts, social density as well as semiprivate spaces (e.g., Fleming, Baum, & Singer, 1985). To date, there are only a few studies that explicitly focus their analysis of context on universities (e.g., Hays & Oxley, 1986; Lakey, 1989; Moos, 1979; Perl & Trickett, 1988).   Unfortunately, the validity of these studies is often affected by the small sample size as well as by a certain arbitrary selection of indicators to measure context.   However, the literature about questions of social climate and learning culture at universities allows us to develop the hypothesis that, along with the learning culture, the architectural form of campus rooms, the number of students, and the gender distribution within an area of study also play an important role (Huber, 1991).

The questions discussed in this study were based on a hierarchical structure; that means we analyzed units grouped at different levels.   On the first level, each student was recorded with his or her individual characteristics.   On the second level, we recorded the context, which was understood in terms of characteristics relevant to the area of study. In accordance with geographical and social psychological literature on the relationship of place and health, the term "context" or "contextual variable" is used here for the characteristics of the area of study ("true" ecological variables independent of subjects' perceptions as well as aggregated subject-level variables like average perceived academic pressure).

In order to analyze these hierarchically structured data appropriately, we used multilevel modeling (Bryk & Raudenbush, 1992; Goldstein, 1995). This statistical method has been used mainly for research on education, for example, to examine the influence of individual, class, and school characteristics on achievement of pupils (Goldstein, 1987).   It has also been utilized to analyze the importance of individual characteristics with regard to regional patterns in health related behavior, as well as voting behavior (Duncan, Jones, & Moon, 1998).   However, it has rarely been used for the analysis of social relationships. Exceptions are studies on different kinds of social networks (e.g., Snijders, Spreen, & Zwaagstra, 1995) as well as on attachment change in newlyweds (Davila, Karney, & Bradbury, 1999). Because the current study of the development of social resources using multilevel modeling cannot rely on existing empirical knowledge, it contains a certain heuristic element.

Questions of the present study

This chapter investigates the conditions of the development of social resources in a new social environment; in this case, the start of university studies. In the latest research on relationships, creation and use of social resources is conceptualized as a transactional process between individuals and their environment. As analyses have shown so far (Bachmann, 1998), students at the two universities (University of Zurich and Federal Institute of Technology) involved in this study differed significantly, depending on the area of study, in terms of number and type of contacts they created with their classmates during the first year of studies.

Therefore, the following questions are placed at the center of these analyses:

1. Which individual predictors promote the creation of social resources in the areas of study (level 1)?

2. Can differences between the areas of study be noticed even when the personal characteristics of the students (compositional effect) are statistically controlled (level 2)?

3. Which aspects of the context — here the area of study — can possibly explain the residual group level variance (level 2)?

METHOD

Procedure

A longitudinal study of the first year of studies was carried out and data were collected at two points in time; before the start of studies (first questionnaire) and again, 10 months later (second questionnaire) to assess changes that might have occurred during the course of time. Every incoming student of the two universities in Zurich (University of Zurich and the Federal Institute of Technology, Zurich) was first surveyed a few weeks before the start of the 1994 autumn term. (Student contact information was obtained from registration information at the universities.) The students were given a self-administered questionnaire, that they were asked to complete anonymously. However, they were asked to add a personal code to the questionnaire so that the subsequent questionnaire could be linked to the first one. The second survey, also using a written questionnaire, was administered in the summer of 1995, a few weeks before the end of the first year.

Sampling and sample

Of the 4,247 students who registered in 1994 for studies at the two universities in Zurich, 3,066 (72%) completed and returned the first questionnaire. The second questionnaire was answered by 1,992 students, resulting in an overall response rate of 47%. In 1,382 cases, the individual code could be linked, resulting in a participation rate of 33%. This sample was found to be representative of the university population in terms of age, gender, and areas of study. This data set is described in detail by Bachmann et al. (1999). A few participants had to be excluded for methodical reasons, because their area of study did not have enough units at level 1 to be included in multilevel analyses. This left us with 1,318 people (level 1) nested within 32 different areas of study (level 2). Per subject, the number of students surveyed ranged from 7 to 143, depending on the number of students enrolled in an area. In simulation studies, a sufficient number of units on level 2 was found to be more relevant than on level 1. When the number of groups was large enough, the estimation of the parameters were sufficiently reliable, even when groups of only 10 students were present (Kreft, 1996).

The study sample was 55% male and 45% female. At the time of the study, the average age of participants was 21.5 years, with a range from a minimum 18 years to a maximum of 65 years. Women comprised 55% of the sample at the University of Zurich and 31% of the sample at the Federal Institute of Technology.

Operationalization

The individual and contextual characteristics that have been identified in the literature as having an influence on the creation and availability of social nets and resources were mentioned earlier. Table 8.1 shows the four response variables that encompass different parameters of social resources, as well as the individual and contextual or group-level predictors.

The response variables were gathered toward the end of the academic year, which constituted the second point of measurement. Three items describe the structural dimensions of social resources, that is, the number of superficial and close relationship networks within an area of study. A further variable, describing the degree of satisfaction with the support received from fellow students within the same area of study was also included. The predictors were assigned to the individual (level 1) or the

TABLE 8.1
Response Variables, Individual and Contextual Predictors

Kind of Variables	Names	Operationalization
Level 1 Responses:	superficial relationships	N of fellow students the focal person had a conversation with during the last 2 weeks
	companionship	N of fellow students with whom the focal person has spent his spare time with
	close relations	N of fellow students with whom the focal person has had a confidential talk
	social support satisfaction	perceived satisfaction of focal person with support received from and relationships with fellow students (6 items, Sommer & Fydrich, 1991)
Level 1 Predictors:	gender: female (FEM)	variable with contrast coding (0=male; 1=female)
	age (AGE)	years of age
	empathy (EMP)	6 items, range 1–5, Becker (1989)
	self-esteem (SEST)	4 items, range 1–5, Rosenberg (1979)
	network-orientation (including family attachment (FAM); self-disclosure (DISC), readiness to create contact (CONT), readiness to search for help (HELP)	10 items, range 1–5, Röhrle (1994)
Level 2 Predictors:	size of area (SIZE)	N of students enrolled in an area of study
	time spend together in the courses (TIME)	N of required lessons in an area of study
	gender distribution (GD) * sex	proportion of male students enrolled in an area of study (2 interaction parameters*sex)
	perceived academic pressure (PRESS)	mean of perceived academic pressure in an area of study (range 1–4, with 1 = deep pressure)

contextual area (level 2). Individual predictors were recorded at the first measuring point so that the student's characteristics could be registered before the start of their studies. As mentioned earlier, in research literature, a multitude of personal characteristics are discussed as connected with social relationships. The seven predictors included in this study have shown consistent effects on the development of social resources (Bachmann, 1998; Hansson et al., 1984; Lakey et al., 1996). For pragmatic reasons, neither the architectural–spatial conditions (e.g., availability of semiprivate learning spaces) nor the structure of the courses (e.g., proportion of group exercises vs. lectures) could be recorded within the framework of the present study. However, it was possible to include the following contextual conditions that are more easily recorded and that, according to the literature, are relevant to the questions raised in this study: The number of students enrolled in an area of study, which varied from less than 20 who were majoring in philosophy and theology to nearly 500 in the "mass" majors such as law; The number of required lectures, which represented the amount of time students had to spend together in the courses, with values ranging from 7 hours per week in the social science and arts, to 40 hours in architecture; Gender distribution with percentage of male students comprising anywhere from 12% to 96% of a subgroup; The overall perceptions of academic pressure or fear of being excluded from a chosen major because of failing the examinations. This was measured by a scale ranging from 1–4, where 1 represented low pressure and 4 represented high pressure. Higher values were found in computer sciences and medicine with 3.3 and particularly low values in theology and sociology with 1.3. For more detailed informations about the context characteristics of the 32 areas of study, see Table 8.2.

Because data (Turner, 1994) suggested that a high proportion of male students has a different impact on the social resources of female students than on those of other male students, we included gender distribution in the equation as two separate interaction parameters for each gender. Furthermore, it is worthwhile noting that three of the contextual predictors were objective parameters provided by the university administration (i.e., number of people enrolled in a particular area of study, the required number of courses within a major, and gender distribution). In contrast, collective perception of academic pressure was used as an aggregate value: the mean of perceived academic pressure within an area of study among the students surveyed.

Statistical procedures

The comparison of individual effects (level 1) with the environmental influence of the areas of study (level 2) required a method that

TABLE 8.2

Contextual Characteristics of the 32 Areas of Study

Area of Study	N Sample	Size	Time	GD	Press
Theology	7	18	10	44	1.29
Philosophy	9	19	9	63	1.67
Psychology	60	137	9	21	3.37
Education	11	32	14	28	1.27
Sociology	22	49	9	49	1.77
German	45	113	8	37	1.49
French	18	34	8	23	1.72
English	35	84	9	30	1.43
History	41	129	7	75	1.76
Ethnology	11	41	12	12	1.36
Law	143	497	16	51	3.12
Economy	93	319	22	81	2.91
Medicine	114	353	25	53	3.32
Veterinary Medicine	24	60	26	20	3.54
ChemistryU	10	21	19	62	2.30
Geology	26	77	20	77	2.08
BiologyU	41	89	25	47	2.39
Architecture	70	285	40	60	2.99
Process Engineering	29	116	25	87	2.55
Mechanical Engineering	50	174	29	96	3.30
Electrical Engineering	36	185	27	96	3.11
Computer Science	33	119	24	94	3.33
ChemistryT	24	59	29	87	3.00
Pharmacy	32	71	31	16	3.16
Forest Sciences	15	53	31	85	2.73
Agricultural Sciences	25	94	32	53	2.64
Civil Engineering	22	79	23	83	2.14
Mathematics	46	171	25	85	2.85
BiologyT	36	103	30	54	2.92
Environmental Sciences	36	79	26	68	2.06
Earth Sciences	20	53	24	75	2.40
Physical Education	35	136	20	59	2.74

Note. Study areas existing at both universities are specified
with U (University) and T (Federal Institute for Technology)

simultaneously handled both levels of factors. The best suited method for such a purpose was multilevel modeling (for an introduction to the methodological principles of multilevel modeling, see the other chapters in this volume). The analyses were carried out using the statistical program MLn (Rasbash, Yang, Woodhouse, & Goldstein, 1995). For the sake of simplicity, only linear functions were postulated. To make the interpretation of the intercept easier, the predictors on level 1 were centered around the mean of the whole sample. In this way, the intercept could be interpreted as the predicted value of a student with average personal characteristics (e.g., student at average age instead of student at age 0, which would be outside the range of the observed values). For parameter estimation in model construction, we used the "iterative generalized least squares" (IGLS) method, an adequate procedure for continuosly distributed response variables when the residuals have normal distributions (Goldstein, 1986; Goldstein, 1995, p. 22). As Raudenbush (1994), showed IGLS is formally equivalent to the procedure VARCL developed by Longford (1987). In a converging model, the parameter estimations are equivalent to the maximum likelihood criteria (Goldstein, 1995, p. 22). The fit check of a model is made by comparing the likelihood ratios (-2 log-likelihood) of a simpler and a more complex model. If the sample is large enough, the differences of likelihoods between the models have an approximate distribution of χ^2. The differences in the number of parameters within each of the two models represent the degrees of freedom when checking the significance of the fit improvement provided by the more complex model (Duncan et al., 1995, p. 68; Goldstein, 1995, p. 32). Testing for the significance of single fixed and random parameters is made by a comparison between the parameter estimation and the standard error. The reader can use the following rule of thumb as a rough guideline: A parameter reaches a significant explanatory value ($p = .05$) when the variance shown is twice as large as the standard error. Testing for significance is made with a χ^2 test. For each of the four response variables, we proceeded to form our model in the following way.

First, we tested a simple Model A with random intercept on level 2. All the relevant individual characteristics were included in this model. The first thing we saw on level 1 of the model (variance among the individuals), was which individual properties influence the response variable. On level 2 (variance between areas of study), the intercept was allowed to vary. This made it possible to find out if there was a contextual effect when the individual characteristics of the students were controlled. By proceeding in this manner, we could compare the compositional and the contextual effect.

The next step (Model B) was an attempt to explain the significant variances between the areas of study by contextual variables. The effect of the different contextual variables (level 2) was analyzed by stepwise inclusion in the equation and was tested by calculating the likelihood ratio test criterion (referring to a χ^2 Table) to see if they significantly improved the explanatory value of the model.

The two statistical models used, simple random intercept models without and with independent contextual variables, could be described by two equations with the two levels i (individual) and j (area of study) explaining, for example, social support satisfaction by level 1 predictors age and self-esteem, and by level 2 predictor size of area of study (number of students enrolled in an area). Equation 8.1 represents Model A without contextual variables:

$$Y_{ij} = \beta_0 X_{0ij} + \beta_1 X_{1ij} + \beta_2 X_{2ij} + (\mu_{0j} + \epsilon_{ij}) \qquad (8.1)$$

where Y_{ij} = satisfaction with social support for student i in area j, X_{0ij} = base category (a set of 1's), and X_1 = age (centered around mean), X_2 = self-esteem (centered around mean).

The fixed parameters could be interpreted as; β_0 = the satisfaction with social support for individuals of average age and self-esteem, β_1 = linear increase of satisfaction with age, and β_2 = linear increase of satisfaction with self-esteem.

The random terms were; ϵ_{ij} = residual value of the individual i within the area j (residuals on ϵ_{ij} level 1), μ_{0j} = differential intercept of area j (residuals on level 2).

The between-individuals and between-areas differences were estimated by the following variance terms; σ_ϵ^2 = level 1 variance (between individuals), and $\sigma_{\mu_0}^2$ = level 2 variance (between areas).

In Model B represented in Equation 8.2 we included contextual predictors (e.g., size of area of study):

$$Y_{ij} = \beta_0 X_{0ij} + \beta_1 X_{1ij} + \beta_2 X_{2ij} + \gamma_1 \omega_{1j} + (\mu_{0j} + \epsilon_{ij}) \qquad (8.2)$$

where γ_1 = linear relationship between area differences and contextual variable ω_{1j}, where ω_{1j} is the contextual variable (size of area).

RESULTS

Before presenting the multivariate analyses, the descriptive measures of the response variables are briefly discussed (see Table 8.3). They describe the size of the social networks and social support satisfaction that emerged during the first year of studies. On average, students declared having a superficial contact with 11 fellow students, having already spent their spare time with about six students and having talked about personal, intimate

TABLE 8.3

Descriptive Measures of the Response Variables at Level 1

Response Variables	N	Mean	SD	Median	Min.	Max.
Superficial relationships	1290	10.7	8.9	9	0	45
Companionship	1290	6.2	6.8	4	0	35
Close relationships	1290	3.2	2.9	2	0	15
Social support satisfaction	1300	3.07	0.52	3.2	1.2	4

topics with three students. With a mean of about 3 (on a scale ranging from 1 to 4), they reported quite a high degree of satisfaction with the social support they received.

We used IGLS as the estimation procedure to analyze these data. In the following, we present the four multivariate models. In the first stage, however, they were a calculated only with the individual predictors and a random intercept (Model A). If a significant effect was observed on level 2, we explained this variance between the areas of study by contextual predictors in a second model (B).

Superficial relationships

The left half of Table 8.4 displays the results of the analysis of the development of the network constituted by superficial relationships. The multivariate model with individual predictors and random intercept (Model A) shows that the average student (male, 21 years old, "average" personal characteristics) engaged in superficial relationship with 10 other students. Gender and age effects were noticeable with women having approximately two people more in their network, which was a significant difference. The older the students were, the fewer superficial relationships they had (with other students). We also found a tenuous link to self-esteem and family attachment. The remaining variance between the areas of study was highly significant, with an intra-area correlation (IAC) of 28%. In order to visualize the differences between the areas of study, we show the differential intercepts of Model A in Fig. 8.1.

Students majoring in physical education had the largest network of superficial contact persons $(10 + 17 = 27)$, whereas those majoring in philosophy and sociology had the smallest network $(10 - 5 = 5$ persons). Because the subject of sports showed an exceptionally high value, we checked whether the results of this model were distorted by this outlier unit. For this purpose, a further model was calculated to study the effect

TABLE 8.4
Multilevel Estimates for Models of Superficial Relations and Companionship

| | Superficial Relationships | | | |
| | Model A | | Model B | |
Parameter	Estimate	(SE)	Estimate	(SE)
Fixed effects				
Level 1				
Intercept	9.66		11.08	
FEM	1.88	(0.50)	-2.76	(1.54)
AGE	-0.18	(0.06)	-0.18	(0.06)
EMP	0.81	(0.70)	0.88	(0.70)
SEST	0.80	(0.41)	0.81	(0.41)
FAM	0.98	(0.42)	0.95	(0.42)
DISC	-0.18	(0.44)	-0.13	(0.44)
CONT	0.65	(0.43)	0.69	(0.43)
HELP	-0.20	(0.37)	-0.23	(0.37)
Level 2				
SIZE			0.00	(0.01)
TIME			0.30	(0.09)
GD*M			-0.02	(0.03)
GD*F			0.07	(0.03)
PRESS			-0.35	(1.59)
Random effects variance				
Level 1 $\sigma^2_{\epsilon_0}$	56.56	(2.32)	56.07	(2.30)
Level 2 $\sigma^2_{\mu_0}$	21.88	(6.02)	13.66	(3.94)
IAC	0.28		0.08	
	Companionship			
Fixed effects				
Level 1				
Intercept	5.92		5.85	
FEM	0.62	(0.40)	0.72	(0.40)
AGE	-0.26	(0.05)	-0.26	(0.05)
EMP	0.32	(0.57)	0.38	(0.57)
SEST	0.38	(0.33)	0.37	(0.33)
FAM	0.39	(0.34)	0.36	(0.34)
DISC	0.64	(0.36)	0.68	(0.35)
CONT	0.46	(0.35)	0.45	(0.35)
HELP	-0.05	(0.30)	-0.03	(0.30)
Level 2				
SIZE			0.00	(0.00)
TIME			0.20	(0.05)
GD*M			-0.01	(0.02)
GD*F			0.03	(0.02)
PRESS			-0.79	(0.87)
Random effects variance				
Level 1 $\sigma^2_{\epsilon_0}$	37.3	(1.53)	37.31	(1.53)
Level 2 $\sigma^2_{\mu_0}$	6.59	(2.00)	3.83	1.29
IAC	0.15		0.09	

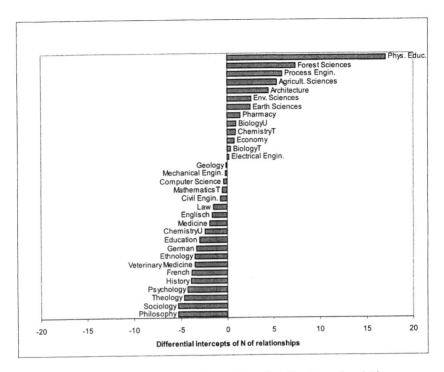

FIG. 8.1. Superficial relationships (residuals on level 2).

of removing the unit of physical education students from level 2 variation by fitting a separate constant (a dummy variable with 1 = physical education as major; 0 = other students) in the fixed part of the Model A. The variance on level 2 was indeed clearly reduced $[\sigma_{\mu_0}^2 = 9.81$ (3.04), $\chi^2 = 10.7$, $p = .001$; intraarea correlation= 0.15], but still significant. The fixed parameters were only very slightly changed. Thus, based on these results, the students of physical education were reintegrated into the model.

In a second stage, we tried to explain the variance on level 2 by contextual predictors. As Model B shows, through this method, the variance between the areas of study was clearly reduced to 8%. The amount of time students spend together in the classroom has a strong positive effect on the creation of superficial relationships. Including this parameter produced a likelihood ratio test criterion of -11 ($df = 2$), which was significant. Neither the size of the area of study (number of students), nor the collective perceptions of academic pressure had any influence on the development of these social resources.

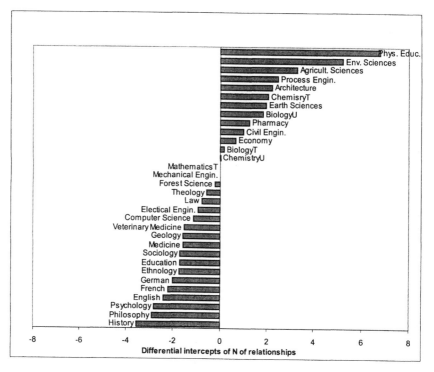

FIG. 8.2. Companionship (residuals on level 2).

Companionship

On average, the students spent their spare time with six different people within their area of study (see right half of Table 8.4). The older the student, the fewer the relationships in which he or she engaged during the first year of studies. No further significant relationship to individual predictors was observed. The contextual effect was also significant in the field of companionship. With 15%, however, the proportion of variance on level 2 (between areas) was clearly smaller than in superficial relationships. Figure 8.2 shows the 32 differential intercepts of the areas of study. In this model, physical education again created the largest network (15 people). Students in environmental sciences, agricultural sciences, civil engineering, and architecture also had large networks. The areas of study with particularly small networks were in the field of arts and social sciences, especially history, philosophy, psychology, English, and French.

Because the differential intercept of the physical education student group was very large, we proceeded the same way as with superficial relationships, and checked to what extent this group influenced the results of the model. By fitting a separate constant for this group, the level 2 variance was reduced [$\sigma_{\mu_0}^2 = 4.38$ (1.4), $\chi^2 = 9.28$, $p = .002$], but still remained significant. The intraarea correlation was reduced from 0.15 to 0.11. The fixed parameters were unchanged, so that again, this group could be reintegrated in the normal sample.

Including the contextual variables in Model B reduced the level 2 variance from 15% to 9.3%. In this model too, the time spent together in the classroom had an influence on the development of social resources. The likelihood ratio test criterion was -14 ($df = 1$), which was highly significant. Neither the number of students in a particular area of study, nor gender distribution within a study area, nor the collective perceptions of academic pressure on the part of the students had any influence on the size of companionship networks created within an area of study.

Close relationships

As Table 8.5 shows (in its left half), the size of close relationship networks that are created during studies can only be predicted on the individual level (Model A) by self-disclosure. This result is not surprising, as it appears plausible that this personal characteristic is a prerequisite condition to building close and trustful relationships. What might be surprising, however, is the absence of gender effect, although the literature gives us clear indications that women have better networks, particularly with respect to intimate relationships (e.g., Bell, 1991; Cramer, Riley, & Kiger, 1991).

In this model, too, the contextual effect was significant but clearly weaker than with the two looser forms of networks. Intraarea correlation (IAC) of this model accounted for only 5.7% of the total variance. The residual values of the single areas of study can be found in Fig. 8.3.

History students had the smallest networks, which consisted of two people. They were followed by students with other majors in the arts and social sciences; philosophy, German, sociology, and psychology. Again, we found the largest networks among physical education majors (with a rate of 4.4 persons). This was also the case with students majoring in pharmacy, agricultural sciences, forest sciences, and veterinary medicine. As the physical education majors had a very high residual value again, we checked the particular effect of this group on the validity of this model. When this group was removed from the level 2 variance and integrated as a fixed parameter into the model, as expected, the group-level variance dropped, $\sigma_{\mu_0}^2 = 0.35$ (0.15), but remained significant ($\chi^2 = 5.24, p = .021$). An important aspect is that the results were unchanged with respect to the

TABLE 8.5

Multilevel Estimates for Models of Close Relationships and Support Satisfaction

| | Close Relationships | | | |
| | Model A | | Model B | |
Parameter	Estimate	(SE)	Estimate	(SE)
Fixed effects				
Level 1				
Intercept	3.09		4.11	
FEM	0.32	(0.19)	-0.50	(0.55)
AGE	-0.03	(0.02)	-0.03	(0.02)
EMP	0.40	(0.26)	0.44	(0.27)
SEST	0.21	(0.15)	0.21	(0.15)
FAM	0.04	(0.15)	0.04	(0.16)
DISC	0.37	(0.17)	0.39	(0.17)
CONT	0.01	(0.14)	0.01	(0.16)
HELP	0.01	(0.14)	0.01	(0.14)
Level 2				
SIZE			0.00	(0.00)
TIME			0.08	(0.02)
GD*M			-0.02	(0.01)
GD*F			0.00	(0.01)
PRESS			0.26	(0.27)
Random effects variance				
Level 1 $\sigma^2_{\epsilon_0}$	8.19	(0.33)	8.15	(0.33)
Level 2 $\sigma^2_{\mu_0}$	0.50	(0.19)	0.21	(0.11)
IAC	0.06		0.025	
	Social Support Satisfaction			
Fixed effects				
Level 1				
Intercept	3.05		3.09	
FEM	0.08	(0.03)	0.09	(0.03)
AGE	0.00	(0.00)	0.00	(0.00)
EMP	0.11	(0.04)	0.12	(0.04)
SEST	0.05	(0.03)	0.05	(0.03)
FAM	0.09	(0.03)	0.09	(0.03)
DISC	-0.02	(0.03)	-0.02	(0.03)
CONT	0.16	(0.03)	0.16	(0.03)
HELP	0.06	(0.02)	0.06	(0.02)
Level 2				
SIZE			0.00	(0.00)
TIME			0.01	(0.00)
GD*M			0.00	(0.00)
GD*F			0.00	(0.00)
PRESS			0.04	(0.04)
Random effects variance				
Level 1 $\sigma^2_{\epsilon_0}$	0.23	(0.01)	0.23	(0.01)
Level 2 $\sigma^2_{\mu_0}$	0.01	(0.00)	0.01	(0.00)
IAC	0.04		0.025	

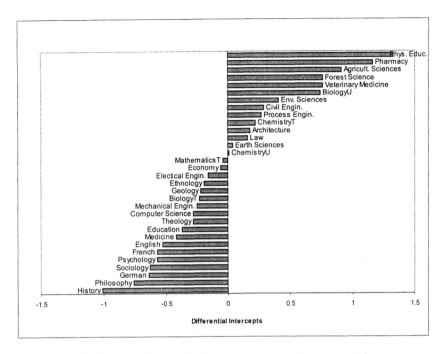

FIG. 8.3. Close relationships (residuals on level 2)

fixed parameters.

Model B reveals the following relationships: The time students had to spend together reduced the level 2 variance by more than half. The likelihood ratio test criterion was -17 $(df = 1)$, and this was highly significant. Gender distribution within the area of study had an influence only on male students: The higher the proportion of male students in an area of study, the smaller their network of closer relationships. In contrast, no equivalent contrary effect could be observed among female students in such a situation. When these two interaction terms were included, the fit of the model changed only slightly. The likelihood ratio test criterion was -6 $(df = 2)$. The two contextual variables time and proportion of males allowed us to almost completely explain the remaining variance between the areas of study. The intraarea correlation of 2.5% was no longer significant.

Satisfaction with social support

Satisfaction with social support is the result of a subjective comparison between available resources and one's needs in this regard. It is another aspect of social resources that might be more closely related to personal characteristics, expectations, and so forth, than the structural parameters presented so far, that is, the number of relationships. In Model A (see Table 8.5, right half), we found that satisfaction with social support was actually more influenced by individual factors. Nevertheless, in the model with individual predictors and random intercept, this dimension of social resources also showed a substantial contextual effect, which constituted 4.2% of the total variance. We found five individual characteristics presenting a relationship to the response variable: Women were more satisfied with the social support they received than were the men; the stronger the empathy, family attachment, readiness to engage in contact, and readiness to seek other people's help, the greater was the resulting satisfaction with support from fellow students. The differential intercepts on level 2 of this model are shown in Fig. 8.4: The subjects with the lowest satisfaction were again from the area of arts and social sciences (philosophy, history, German, sociology) with the exception of architecture students who also reported a low degree of satisfaction. The highest degree of satisfaction was found in physical education, computer science, agricultural sciences, environmental engineering, and forest sciences.

Model B had only one additional predictor that made a significant contribution. Again, the contextual characteristic that influenced satisfaction was the subject's required number of courses: The more hours a week students had to spend together, the higher the satisfaction with social support (likelihood ratio test criterion was -11, $df = 1$). By this contextual effect, group-level variance was reduced to 2.5%.

DISCUSSION

In this study about the development of four different dimensions of social resources during university studies, we found consistent contextual effects that go beyond the compositional characteristics. Students who were in the arts or social sciences — regardless of age, gender, self-esteem, empathy, or readiness to engage in social contact — had more difficulties building a social network big enough to satisfy their needs. The variance between the areas of study changed depending on the dimension of resources: The size of rather superficial relationship networks was more strongly influenced by the context, than was the size of closer relationship networks. A sufficient number of superficial contacts has been shown to be particularly important for coping with daily hassles as well as for developing a feeling of integration and identification toward a social group (Bachmann et al., 1999; Rook, 1987). The individual predictors we have examined in our study do not show a consistent effect through all four models; depending on the kind of relationship or the resource dimension, we found different

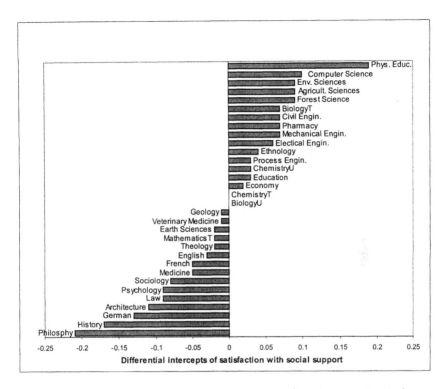

FIG. 8.4. Satisfaction with social support (residuals on level 2)

personal characteristics to be relevant for building social contacts. Age had an important influence on rather superficial relationship networks; the older the student, the smaller the network. However, older students were not less happy with their relationships and the support they received, which indicates that these smaller networks did meet their needs. Our findings regarding gender in the university setting suggest it to be lower in importance in comparison to what has been previously reported in the literature. Female students, however, clearly reported a higher degree of satisfaction with their relationships and the support received than did male students. The stepwise inclusion of contextual variables in the models allowed a substantial reduction of variance between the areas of study. By this, we were able to show that the level 2 variance noticed in the simpler model is indeed due to the ecological conditions. In particular, the time that students spend together in the classroom can explain a significant part of the variance between the areas of study and it is the only predictor at all that shows a consistent effect on all four dimensions of social resources.

A possible explanation for this result could be found in the fact that a certain amount of required courses per subject were necessary to generate a kind of belongingness among the students. We also found, in certain cases, an effect produced by gender distribution within the areas of study: The higher the proportion of males within a subject area, the larger the number of superficial relationships of female students, and the smaller the number of close relationships of male students.

The explorative nature of this study carries a certain number of theoretical and methodological problems. For pragmatic reasons, we had to limit the choice of contextual predictors to those that were easily accessible. We believe that by having done so, essential contextual characteristics, such as the architectural environment, were disregarded. The dynamic interaction process between the individual and his or her context, as it is postulated in the transactional theory, could only be represented as a simplified model. In order to analyze transactional processes in a more appropriate way, it would be necessary to choose a design that contains much more than two points of measurement, and that is one that allows the modeling of feedback loops between an individual and his or her context (cross-level transactions). In this study, we could only analyze the effect of personal characteristics measured before the beginning of the studies on the outcome of size and quality of social resources after 10 months of studies in a specific context. This research design did not allow the analysis of the reciprocal effect (influence of the context on the personal characteristics of the students).

Finally, we have to address the problem of dealing with very different numbers of units per subject area. In multilevel analysis, a small number of level 1 units in a level 2 unit is regarded as a relative lack of information about this special group so that the best estimate of the predicted residual should be close to the overall population value (under the assumption that this group belongs to a population of units). As the number of level 1 units in a group decreases, the "shrinkage factor" (ranging from 0 to 1) becomes closer to zero (Goldstein, 1995, p. 24) so that the predicted residual value gets closer to the grand mean. Thus, the particularities of the smaller areas of study were systematically brought into approximation to those of the large areas of study.

Despite these problems, which indicate where future studies could be improved, the results presented in this study also demonstrate that the use of multilevel modeling is an appropriate statistical method to analyze how resources originate in a determined context, which is a question of increasing importance in the health sciences. The method allows new and very promising insights about the interaction of an individual and his or her context. Furthermore, practical implications for prevention and health promotion in university settings can also be derived from the results we obtained. Social resources can be promoted through the structuring of time that students spend together in the classroom: Students from areas of study with a smaller number of commonly required courses had smaller social networks; they were less satisfied with the support they received and therefore were more prone to experience stress in this transitional phase.

SUMMARY

In this study, we analyzed personal and contextual determinants of the development of social resources among students in the first year of their university studies. According to the ecological approach, social resources are defined as multidimensional constructs that are determined by transactions between a person and his or her environment, and occurring within a specific ecological context. Therefore, according to this conceptualization, it is necessary to consider both the individual and his or her specific context; it is also important to simultaneously analyze the effect of individual and contextual variables in one model. The following three questions were answered in the present chapter: 1. Which personal characteristics are crucial in the formation of different social resource dimensions?

2. Does context exert an effect that transcends the influence of personal qualities?

3. If yes, can the level 2 variance be explained by context-related variables?

Level 1 predictors consisted of gender, age, self-esteem, empathy, and network, orientation. Level 2 predictors consisted of the size, time spent together, gender distribution and perceived academic pressure in the area of study. Response variables were comprised of the number of superficial relationships, companionship, and close relationships, as well as perceived satisfaction with social support. We tested four different multilevel models to explain the formation of the different dimensions of social resources in a sample of 1,318 students (level 1), nested within 32 areas of study (level 2). Depending on the resource dimension, we found different personal characteristics to be relevant for building social contacts. A significant contextual effect was observed in all four models. The largest group-level effect was seen within networks consisting of superficial relationships. A considerable part of the variance between the study areas was explained by the number of required lessons (i.e., time spent together in the classroom). This predictor was the only one to have a consistent effect on all four dimensions of social resources. Gender distribution in an area was found to have an effect on the development of both superficial and close networks.

Contrary to the subjective perception of students and the assumptions of student counselors, the size of the student group as well as perceived academic pressure seem not to have any substantial influence on the development of social resources among students.

ACKNOWLEDGMENT

We wish to thank the following people and institutions: Min Yang from Multilevel Models Project, Institute of Education, University of London for methodological support, Elmar Brockmann for technical support and the heads of the academic departments of the University of Zurich and the Federal Institute for Technology, Zurich for financial support.

References

Bachmann, N. (1998). Die Entstehung von sozialen Ressourcen abhängig von Individuum und Kontext. Ergebnisse einer Multilevel-Analyse. Münster: Waxmann.

Bachmann, N., Berta, D., Eggli, P., & Hornung, R. (1999). Macht Studieren krank? Die Bedeutung von Belastung und Ressourcen für die Gesundheit der Studierenden. Bern: Huber.

Barrera, M., & Baca, L. M. (1990). Recipient reactions to social support: Contributions of enacted support, conflicted support and network orientation. *Journal of Social and Personal Relationships, 7,* 541–551.

Becker, P. (1989). Der Trierer Persönlichkeitsfragebogen. Handanweisung. Göttingen: Hogrefe.

Bell, R. A. (1991). Gender, friendship network density, and loneliness. *Journal of Social Behavior and Personality, 6,* 45–56.

Bryk, A. S., & Raudenbush, S. W. (1992). *Hierarchical linear models: Applications and data analysis methods.* London: Sage.

Burda, P. C., Vaux, A., & Schill, T. (1984). Social support resources: Variation across sex and sex role. *Personality and Social Psychology Bulletin, 10,* 119– 126.

Cramer, L. A., Riley, P. J., & Kiger, G. (1991). Support and antagonism in social networks: Effects of community and gender. *Journal of Social Behavior and Personality, 6,* 991–1005.

Davila, J., Karney, B. R., & Bradbury, T. N. (1999). Attachment change processes in the early years of marriage. *Journal of Personality and Social Psychology, 76,* 783–802.

Duncan, C., Jones, K., & Moon, G. (1998). Context, composition and heterogeneity: Using multilevel models in health research. *Social Science and Medicine, 46,* 97–117.

Fleming, R., Baum, A., & Singer, J. E. (1985). *Social support and the physical environment.* In S. Cohen & S.L. Syme (Eds.), *Social support and health* (pp. 327– 346). New York: Academic Press.

Goldstein, H. (1986). Multilevel mixed linear model analysis using iterative generalised least squares. *Biometrika, 73,* 43–56.

Goldstein, H. (1987). *Multilevel models in educational and social research.* London: Arnold.

Goldstein, H. (1995). *Multilevel statistical models.* London: Edwald Arnold.

Hansson, R. O., Jones, W. H., & Carpenter, B. N. (1984). Relational competence and social support. *Review of Personality and Social Psychology, 5,* 265–284.

Hays, R. B., & Oxley, D. (1986). Social network development and functioning during a life transition. *Journal of Personality and Social Psychology, 50,* 305– 313.

Hobfoll, S. E., & Freedy, J. R. (1990). The availability and effective use of social support. *Journal of Social and Clinical Psychology, 9,* 91–103.

Hornung, R., & Gutscher, H. (1994). Gesundheitspsychologie: Die sozialpsychologische Perspektive. In P. Schwenkmezger & L. R. Schmidt (Eds.), *Lehrbuch der Gesundheitspsychologie* (pp. 65–87). Stuttgart: Enke.

Huber, L. (1991). Sozialisation in der Hochschule. In K. Hurrelmann & D. Ulich (Eds.), *Neues handbuch der sozialisationsforschung* (pp. 417–441). Weinheim: Beltz.

Kreft, I. G. (1996). *Are multilevel techniques necessary? An overview including simulation studies.* http://www.stat.ucla.edu/-kreft.

Lakey, B. (1989). Personal and environmental antecedents of perceived social support developed at college. *American Journal of Community Psychology, 17* (4), 503–519.

Lakey, B., McCabe, K. M., Fisicaro, S. A., & Drew, J. B. (1996). Environmental and personal determinants of support perceptions: Three generalizability studies. *Journal of Personality and Social Psychology, 70* (6), 1270–1280.

Leppin, A., & Schwarzer, R. (1990). Social support and physical health: An updated meta-analysis. In R. Schmidt, P. Schwenkmezger, J. Weinman, & S. Maes (Eds.), *Health psychology: Theoretical and applied aspects* (pp. 185–202). London: Harwood.

Longford, N. T. (1987). A fast scoring algorithm for maximum likelihood estimation in unbalanced mixed models with nested random effects. *Biometrika, 74,* 817–827.

Moos, R. (1979). *Evaluating educational environments: Procedures, methods, findings and policy implications.* San Francisco: Jossey-Bass.

Perl, H. I., & Trickett, E. J. (1988). Social network formation of college freshmen: Personal and environmental determinants. *American Journal of Community Psychology, 16,* 207–224.

Rasbash, J., Yang, M., Woodhouse, G., & Goldstein, H. (1995). MLn: Command reference guide. London: Institute of Education.

Raudenbush, S. W. (1994). *Equivalence of Fisher scoring to iterative generalised least squares in the normal case with application to hierarchical linear models.* Unpublished manuscript.

Röhrle, B. (1994). *Soziale Netzwerke und soziale Unterstützung.* Weinheim: Beltz.

Rook, K. S. (1987). Social support versus companionship: Effects on life stress, loneliness, and evaluations by others. *Journal of Personality and Social Psychology, 52,* 1132–1147.

Rosenberg, M. (1979). *Conceiving the self.* New York: Basic Books.

Snijders, T. A. B., Spreen, M., & Zwaagstra, R. (1995). The use of multilevel modeling for analysing personal networks: Network of cocaine users in an urban area. *Journal of Quantitative Anthropology, 4,* 475–491.

Sommer, G., & Fydrich, T. (1991). Entwicklung und Überprüfung eines Fragebogens zur sozialen Unterstützung (F–SOZU). *Diagnostica, 37,* 160–178.

Tolsdorf, C. C. (1976). Social networks, support and coping: An exploratory study. *Family Process, 15,* 407 – 417.

Vaux, A. (1990). An ecological approach to understanding and facilitating social support. *Journal of Social and Personal Relationships, 7,* 507–518.

9

Ordered Category Responses and Random Effects in Multilevel and Other Complex Structures

Antony Fielding

University of Birmingham, United Kingdom

Many response variables in multilevel models are ordered categories. Although those used here are school grades, other examples arise in diverse application areas. Rating scales provide important examples. When number labels are attached to the categories, it is only their ordering that has any substantive meaning. However, in practice, it is quite common to apply arbitrary scores to the responses and model them as if they were interval scales with higher measurement properties. There are obvious practical advantages in this in that methodology for linear multilevel models is more widely known and software more accessible.

However, apart from the arbitrariness of the implicit scaling assumptions, many objections to this practice from statistical and substantive viewpoints arise in the literature. These are reviewed in Fielding (1999, 2000). Questions arise as to the wisdom of assuming that effects operate linearly and additively on a chosen scale rather than on some other arbitrarily chosen one. Also, even if the scale of a continuous variable underlying the ordered data could be satisfactorily chosen, there is still the fact of grouping the data. It is well known that grouping of responses leads to biased and poor estimation of regression parameters in standard single level models (Stewart, 1983). Simulations of multilevel models by Fielding (2000) have also shown that the impact of grouping on variance component estimates may be even more problematic.

The methodological developments and applications discussed seek alternatives to multilevel models based on scores. They focus on models that directly deal with grouped ordered responses and do not assume a higher level of measurement than their operational construction allows. The models are also extended from multilevel structures to more complex structures involving cross-classified and weighted random effects. They form part of the family of generalized linear mixed models (GLMMs). They extend the construction of generalized linear models for ordered responses (McCullagh & Nelder, 1989) by the incorporation of random effects.

MODELING ORDERED RESPONSES IN MULTILEVEL HIERARCHIES

A two level example: Progress in primary schools

To assist teachers in meeting learning needs, baseline testing of children entering primary school has been set up in Local Education Authorities (LEAs) in England and Wales. Progress can be traced through to statutory Key Stage 1 (KS1) tests taken 2 years later. The data used here is two level with 4,444 children in 114 schools from one large LEA. The response modeled is a national KS1 reading test with a grading of levels into six ordered categories.

In developing a multilevel model, there may be a variety of analytical aims to meet substantive interest in educational progress. Adjusted outcomes controlling for prior ability of intake are of interest in assessing progress. Other individual and school characteristics, although often correlated with ability, may have a net effect on progress and are of interest. Model results may also inform target setting based on known individual characteristics at baseline. As shown later, when ordered category models are used, predictions can be made of probabilities or "chances" of achieving certain KS1 levels, given individual profiles. The estimation of school effects on progress through well-specified multilevel models, controlling for relevant variables, also establishes a sound methodological basis for "value added" indicators. These aims provide background motivation for the methodology and example to be discussed.

The response variable has three levels with a fine grading of KS1 level 2. Sample distribution details are given in rows 1 to 4 of Table 9.1. Seven baseline teacher–assessed test variables in spelling, reading, writing, number, algebra, shape and space, and handling data are used as initial ability controls. These have a 4-point categorized scale. After investigation, however, it was found satisfactory to enter them into models as equal interval scales; standardized with mean zero and unit variance. Table 9.2

TABLE 9.1

Distribution of Response and Approximate Marginal Expectations From
Various Models

	Level of KS1 Reading	0	1	2c	2b	2a	3
1	Category number	1	2	3	4	5	6
2	Sample frequency	160	871	786	738	532	1357
3	Sample %	3.6	19.6	17.7	16.6	12.0	30.5
4	Cumulative %	3.6	23.2	40.9	57.5	69.5	100.0
	Model predictions						
5	AL cumulative. %	2.8	20.7	39.5	57.9	71.0	100.0
6	AL %	2.8	17.9	18.8	18.4	13.1	29.0
7	AP cumulative. %	2.2	20.6	39.5	57.6	70.6	100.0
8	AP %	2.2	18.4	18.9	18.1	13.0	29.4
9	BL cum. % at means of effect variables	1.3	15.7	38.4	63.2	78.9	100
10	BL % at means	1.3	14.4	22.7	24.8	15.7	21.1

gives definitions and summary measures of these and other explanatory
variables used.

The basic two level model formulation for ordered responses

Taking advantage of the order of the response categories it is convenient to
model the cumulative probabilities $\gamma_{ij(s)}$ (s = 1, 2, ..., 5) for the ith child in
the jth school achieving at least category s on the response. By definition,
$\gamma_{ij(6)} = 1$ and is not explicitly considered. An observed response category
can be converted into a multivariate set of binary indicators, and these can
also be cumulated to form a vector $\mathbf{y_{ij}} = \{y_{ij(1)}, y_{ij(2)}, y_{ij(3)}, y_{ij(4)}, y_{ij(5)}\}$
with $\mathrm{E}(\mathbf{y_{ij}}) = \{\gamma_{ij(1)}, \gamma_{ij(2)}, \gamma_{ij(3)}, \gamma_{ij(4)}, \gamma_{ij(5)}\}$. For many situations,
conditional on the probabilities, $\mathbf{y_{ij}}$ will be multinomial and have a
variance–covariance matrix with elements $\gamma_{ij(s)}(1-\gamma_{ij(s')})$ for $s \leq s'$. An
extramultinomial variation parameter ϕ multiplying this matrix will also
be introduced to allow for possible misspecification of the probabilities. It
is worth noting at this stage that multinomial variation depends directly
on expected values (probabilities). There is no separate estimable variance
parameter akin to the individual disturbance in continuous variable
multilevel models. This fact of inseparability of parametric specifications of
expectations and variances is often a source of some confusion in empirical
work.

Completing a basic multilevel model requires specification of the $\gamma_{ij(s)}$in

TABLE 9.2

TABLE 9.2

Definitions of Variables and Summary Statistics in the Models of Table 9.3

Variable	*Description and summary statistics*
Level 1: Pupil	
Gender	1 = male (50.3%); 0 = female (49.7%)
Free school meals	1, Eligible for free school meals (38.3%); 0, Not eligible (61.7%)
Nursery	1, previous nursery education (66.5%); 0, Others (33.5%)
Centred Age	Age in months centred on 86 months at KS1 testing (SD = 3.5)
Ethnic-Language dummies	14 compound categories formed from 10 ethnic groups and 12 first languages Base for dummies is White all languages (59.0%). Only 0.3% of children were White with languages other than English
EthLang2	1 = Afro-Caribbean-English (7.1%); 0 = Others
EthLang3	1 = Afro-Caribbean-not English (0.4%); 0 = Others
EthLang4	1 = Other ethnic groups-English (5.1%); 0 = Others
EthLang5	1 = Pakistani-not English (16.0%); 0 = Others
EthLang6	1 = Indian-Hindi (0.1%); 0 = Others
EthLang7	1 = Indian-Punjabi (5.8%); 0 = Others
EthLang8	1 = Indian-other languages not English (1.0%); 0 = Others
EthLang9	1 = Bangladeshi-not English (4.0%); 0 = Others
EthLang10	1 = Arabic-not English (0.5%); 0 = Others
EthLang11	1 = Chinese-not English (0.2%); 0 = Others
EthLang12	1 = Vietnamese-not English (0.2%); 0 = Others
EthLang13	1 = Mixed race-not English (0.5%); 0 = Others
EthLang14	1 = Other ethnic groups-not English (0.2%); 0 = Others
Level 2: School	
Baseline aggregate	Average of 7 percentages of children at or above level 2 on baselines
FSM context	Percentage of chldren eligible for free school meals

terms of fixed and higher level random effects.   The probabilities are constrained to lie between zero and one. Usually, through a link function, they will be monotonically transformed to $\alpha_{ij(s)} = F^{-1}\left(\gamma_{ij(s)}\right)$ so that $\gamma_{ij(s)} = F\left(\alpha_{ij(s)}\right)$ and $\alpha_{ij(s)}$ range along the whole real line. The latter can then be formed as a linear predictor (LP) in unconstrained ways familiar in continuous variable linear multilevel models. Because $\gamma_{ij(s)} \leq \gamma_{ij(s')}$ for $s < s' = 1, 2, \ldots, 5$, certain forms of both the LP and F (or its inverse the link F^{-1}) are indicated. Any suitable choice of F will be a distribution function of some continuous random variable. A LP of form,

$$\alpha_{ij(s)} = \theta_s - \{\sum_{\ell=1}^{L}\beta_\ell x_{\ell ij} + u_j\} \tag{9.1}$$

is suitable for the example. The x variables are explanatory variables. The random school effect u_j is usually assumed $\sim N\left(0, \sigma_u^2\right)$. The minus sign in the LP is for interpretational convenience because a positive effect on $\gamma_{ij(s)}$ is a negative effect on the ordered response and vice versa. Because,

$$\sum_{\ell=1}^{L}\beta_\ell x_{\ell ij} + u_j \tag{9.2}$$

is invariant for specific observations, the θ_s demarcate the cumulative probabilities for that individual. If $\theta_1 \leq \theta_2 \leq \theta_3 \leq \theta_4 \leq \theta_5$, as is necessary, then since F^{-1} is nondecreasing in its argument, then so will be $\gamma_{ij(s)}$ with s as required. For identification purposes, the x should not include an explicit intercept variable because this will be aliased with the set of θ_s.

Two forms of the link function, probit and logit will be used. These arise as inverses of respectively the standard normal and standard logistic forms for F. Other links have been suggested, and in some areas, the use of the complementary log–log is useful (Goldstein, 1995). The logit link gives

$$\ell n\left[\frac{\gamma_{ij(s)}}{1 - \gamma_{ij(s)}}\right] = \theta_s - \left(\sum_{\ell=1}^{L}\beta_\ell x_{\ell ij} + u_j\right) \tag{9.3}$$

so that effects in the LP operate on a log-odds scale. Alternatively, the useful proportional odds property of this model is shown by

$$\left[\frac{\gamma_{ij(s)}}{1 - \gamma_{ij(s)}}\right] = exp\{\theta_s - \left(\sum_{\ell=1}^{L}\beta_\ell x_{\ell ij} + u_j\right)\}$$

$$= exp\{\theta_s\}exp\{-\left(\sum_{\ell=1}^{L}\beta_\ell x_{\ell ij} + u_j\right)\}. \tag{9.4}$$

As the values of variables or effects change in the LP, and affect the whole response distribution $\gamma_{ij(s)}$, the effect on the set of odds across s is proportional. Differences of log odds and odds ratios for any pair of s and s' are constant across observations. Odds ratios for pairs of observations are also constant across the five values of s. These facts mean some useful interpretations of parameter estimates in the ordered logit model are possible. Later, in Table 9.3, for instance, a base logit model (BL) yields a 0.8 estimate of the baseline number test coefficient that is the net effect on log odds. A standard deviation increase in number (*ceteris paribus*) shifts the entire response distribution upward in such a way that the set of five log-odds are all decreased by 0.8. The odds themselves in the set change proportionately. A decrease of a standard deviation unit in number multiplies each by $\exp(0.8) = 2.23$. Odds ratios for pairs of categories remain unaffected. Changes in the school random effects also operate in this way.

This model presented so far is very basic. However, the methodology can be extended quite readily in ways familiar in general multilevel modelling. More levels can be introduced into the LP. The regression coefficients can also be regarded as random effects at any level and so on. In some situations, there are also advantages in allowing the θ_s to randomly vary over schools separately instead of being consistently shifted by the single random intercept u_j, or to allow the θ_s to interact with explanatory variables (Hedeker & Mermelstein, 1998; Yang, 2000; Yang, Fielding, & Goldstein, in press-a). In logit models in these cases, odds are no longer proportionally changing.

It is sometimes useful, both conceptually and for results interpretation, to postulate the existence of an unmeasured and unobservable continuous latent variable (lv) underlying the response categories. Contiguous intervals on this variable formed by unknown cutpoints $\{-\infty, \theta_1, \theta_2, \theta_3, \theta_4, \theta_5, +\infty\}$ may be supposed to form the observed categories. It may be further supposed that this lv follows a standard continuous response multilevel model

$$lv = \sum_{\ell=1}^{L} \beta_\ell x_{\ell ij} + u_j + e_{ij}.$$

The intercept fixed coefficient may be taken as zero because the location of the lv is arbitrary. An observation in category s or below occurs when $lv < \theta_s$. The scale of the lv is arbitrary and may be fixed by specifying a priori the variance of the level 1 disturbance e_{ij}. Then, if $e_{ij} \sim N(0,1)$, and with Ω^{-1} denoting the inverse of the normal distribution function, the probit model

$$\Omega^{-1}\left(\gamma_{ij(s)}\right) = \theta_s - \left(\sum_{\ell=1}^{L} \beta_\ell x_{\ell ij} + u_j\right) \qquad (9.5)$$

for the cumulative probabilities ensues. The ordered logit model follows if e_{ij} has a standard logistic with variance $\pi^2/3$. Fixing the scale of the unmeasured lv in these convenient ways resolves the unidentified nature of its variance with only ordered categories observed. The assumption of an underlying latent variable is not unduly restrictive and no measurement scale is being imposed. It is arbitrary up to any order preserving monotonic transformation.   The distributional assumptions merely postulate the existence of a variable governed by the assumed multilevel model by which the ordered category models may be interpreted.

Choices of link function can be the subject of some debate (see Fielding, 1999). Often, although not always, the probit and logit give similar results. This is mainly due to the close affinity between the standard normal and logistic distributions, except for discrepancies in the tails. However, in comparing results, it must be recognized that effects, and hence, parameter values are on different scales implicit in the difference between the lv variances of unity and $\pi^2/3$.

Similarly model developments with the same type of model are subject to implicit scale differences. If, for instance, in an ordered probit model we introduce some additional explanatory effects, we might expect the residual variability of the response at the lowest level to change. However, in operation, the response is effectively restandardized. This may be seen through the lv that is rescaled to always have the same standard level 1 variance as the model develops. Model parameter values including higher level variances will also be rescaled appropriately in addition to any other real changes consequent to the extra model features. The two types of change are not easily separable from explicit model results, which give sizes of effects on the link scale rather than on the response directly. In the logit model, for instance, although changes in effects on log-odds are easy to see, this is not so for the response itself. In standard linear multilevel applications, it is often of interest to see how variance components change at various levels with the introduction of further covariates. It is unfortunate that in empirical work, such interpretations that ignore the implicit response scale changes are often carried over to GLMMs. The later results section outlines an approximate way of handling this matter of scale change.

MODEL ESTIMATION

Many procedures based on likelihood estimation have been suggested (e.g., Anderson & Aitken, 1985; Ezzett & Whitehead, 1991; Harville & Mee, 1984; Jansen, 1990). Most are computationally intensive and as a result, there is a limit to the scope and available software. There is free software MIXOR (Hedeker & Gibbons, 1996) available for some situations. The widely used software HLM (Bryk & Raudenbush, 1992) also contains some facilities for GLMMs. Aitkin (1999) developed the theory of a general maximum likelihood approach implemented in the GLIM4 software but his examples were for binary responses and practical applications on ordered category models have remained relatively untested.

Examples here use the widely available MlwiN software (Rasbash et al., 1999) built around Goldstein's (1995) multilevel iterative generalized least squares (IGLS). Extensive and flexible macro commands and facilities adapt this to a wide range of complex models. For the ordered category models discussed here the macros MULTICAT (Yang et al., 1998) are under continuous development. Goldstein (1995) gave detailed theory of the iterative method. A brief description of the stages is in order here. The elements of the response cumulative indicator vector may be written $y_{ij(s)} = \gamma_{ij(s)} + \delta_{ij(s)}$ ($s = 1, 2, \ldots, 5$). An extra level below the observation level is created with these five elements as units. This specifically caters to the conditional covariance structure of these elements that are updated at each iteration by the latest estimates of $\gamma_{ij(s)}$. The functions

$$\gamma_{ij(s)} = F\left[\theta_s - \left(\sum_{\ell=1}^{L} \beta_\ell x_{\ell ij} + u_j\right)\right], \qquad (9.6)$$

being the conditional expectations of y, are approximated at each stage of an iterative process by a Taylor Series linearization using the latest parameter estimates. By these means, each step casts the structure into the form of a standard multilevel linear model then estimable by IGLS. There are choices. The linearization may be around the fixed part only in which case it becomes equivalent to marginal quasi-likelihood (MQL). Inclusion of first or second order terms in the expansion yield alternatively MQL1 or MQL2. Expansion may also use estimates of random effects $\hat{u}_j$ in which case they penalized (predictive) quasi-likelihood referred to as PQL1 or PQL2. Accumulating evidence has suggested a preference for PQL2, on which all results in this chapter are based. However, there may sometimes be computational and convergence problems increasing in complexity when moving from MQL1 to PQL2. Marginal likelihood is often satisfactory and certainly involves fewer computational problems and time.

The MULTICAT macros also estimate an extramultinomial parameter that can be constrained to unity for imposed multinomial variation. The provenance of extra multinomial variation is usually sought in missing levels or relevant explanatory variables in the model. Evidence has suggested that if the parameter is made free in operation, then this leads to improved model estimation (Yang, 1997). Sparse data is also sometimes a reason claimed for estimates of the parameter different from unity (Fielding & Yang, in press; Wright, 1997). The behavior is an underresearched area so the context must be carefully considered. In the examples to follow, the parameter is left free and its estimated values barely commented on.

THE PRIMARY SCHOOL PROGRESS RESULTS

Results of model fitting are given in Tables 9.3 and 9.4. AL (logit) and AP (probit) are base models exhibiting uncontrolled raw variation in outcome between schools and between children. They establish a framework for model development. Using the inverse of the link function and substituting estimates of θ_s, approximate estimates of the marginal population probabilities for AL amd AP are given in rows 5 to 8 of Table 9.1. The approximation is governed by expectations of nonlinear functions of random variables being only approximately the same as the function of the expectations. Without school random effects, both models would be simple reparameterizations of marginal category probabilities. Both models give similar results and are close to the empirical one. The largest, although still a minor difference, is as might be expected in the very small, lowest KS1 level. The models AL and AP give almost identical results and the different set of estimates of θ_s are brought almost exactly into line once a scale factor of $\pi/\sqrt{3}$ is applied. From the implicit lv level 1 variance in each case, the school contribution to overall response variability is estimated to be almost the same at 15.0% and 16.3%, respectively.

Explanatory models are now developed. Logit model BL and probit BP introduce baseline tests. Using similar comparative criteria as before, they give similar results. Thus, the rest of the results use only the logit. CL uses a range of pupil variables except baseline tests. DL adds these back in. Model EL introduces some of the school context variables found relevant and interesting. It must not be supposed that these model developments were uninformed by substantive and statistical evaluative criteria. Indeed, investigations were quite extensive. Likelihood calculations and hence deviances are unreliable in MlwiN generalized procedures, although for heuristic purposes, they are presented for the models in Table 9.3. For formal evaluation of various model developments, Wald tests of parameters, subsets of them, or contrasts were used (Goldstein, 1995). By these means,

the possibility of baseline coefficients randomly varying across schools
and a wide range of interaction effects were found to be insignificant.
Polynomial terms in baselines that take up ceiling and floor effects in points
scored models (Fielding, 1999) were also not significant. Main effects that
were not statistically significant are presented in Table 9.3. The size of
coefficients, even when small relative to their estimated precision, are often
of substantive interest. Thus, for instance, the nursery variable that was
significant in model CL was not so in DL and EL. This highlights the feature
that a net influence on raw performance ceases to have a real influence on
progress when baseline control is exercised.

Some general comments about the interpretation of the results will now
be made before focusing on two specific practical analytical uses. Because
baseline variables are standardized, the most important of them is seen to
be number. Fixed effect coefficients can be interpreted directly on the scale
of the link transformed cumulative probabilities. However, for instance, the
number coefficient of 0.8 in logit BL can also be interpreted as a net change
in the standard scaled lv for that particular model. The $\gamma_{ij(s)}$ themselves
are nonlinear functions and such marginal net effects on them depend on
values taken by all variables and on random effects in the model. For
summary purposes, partial derivatives of the $\gamma_{ij(s)}$ evaluated at the means
of these values are often useful. Differencing these yield summary effects on
the category probabilities summing to unity. For baseline number, these are
estimated sequentially as -0.010, -0.096, -0.083, 0.003, 0.053. They show
the obvious effect of baseline number test score increases on shifting the
response variable upwards. Rows 9 to 10 of Table 9.1 show the estimated
probabilities from model BL at the similar mean points and school effect of
zero. The greater concentration in the middle categories show the response
variability reduction consequent to the introduction of baseline controls.

The school random effect is also additive on the link or lv scale. In a
particular model and akin to variance component analysis, it is useful to
examine the school variance as percentage of total residual variance using
the standardized lv variance at child level 1. These are presented below the
school variance estimates in Table 9.3. Results for BL and BP are similar
and demonstrate that school variation relative to individual variation is
greater on adjustment for baseline ability than for the raw performance in
AL and AP.

With these ideas fixed, it is of obvious interest to interpret changes in the
models as more effects are introduced across the results in Tables 9.3 and
9.4. As previously discussed, due to artificial implicit scale change effects
on parameters, regardless of other real changes, some extra investigation
is required. For example, as we move from base model AL to introducing
baseline controls in BL, we note from the estimated distributions in Table

TABLE 9.3

Parameter Estimates for Two–Level Ordered Category Models for KS1
Reading Test

	Model AL	Model AP	Model BL	Model BP
Fixed Paramters				
$\theta^{(1)}$	-3.56 (0.11)	-2.00 (0.06)	-4.35 (0.14)	-2.46 (0.08)
$\theta^{(2)}$	-1.34 (0.08)	-0.82 (0.05)	-1.68 (0.12)	-1.00 (0.06)
$\theta^{(3)}$	-0.43 (0.08)	-0.27 (0.05)	-0.47 (0.11)	-0.28 (0.06)
$\theta^{(4)}$	0.32 (0.08)	0.19 (0.05)	0.54 (0.11)	0.30 (0.08)
$\theta^{(5)}$	0.90 (0.08)	0.54 (0.05)	1.32 (0.11)	0.76 (0.06)
Baseline tests				
Spelling			0.21 (0.04)	0.12 (0.03)
Reading			0.35 (0.04)	0.20 (0.03)
Writing			0.21 (0.04)	0.13 (0.03)
Number			0.80 (0.04)	0.46 (0.03)
Algebra			0.32 (0.04)	0.18 (0.03)
Shape & space			0.02 (0.04)	0.01 (0.03)
Handling data			0.28 (0.04)	0.16 (0.03)

Random parameters
School variance: $\hat{\sigma}_u^2$

	0.581 (0.086)	0.201 (0.031)	1.233 (0.177)	0.387 (0.058)
School % of residual variance in lv model	15.0	16.3	27.2	27.9
Approximate reduction in level 1 logistic latent variance from model AL	–	–	72%	–
Approximate rescaled school variance using AL model logistic CMS scale	0.581	–	0.880	–
Extra-multinomial $\hat{\phi}$	0.968 (0.010)	0.988 (0.009)	0.956 (0.009)	1.091 (0.008)
-2 log-likelihood	7680.70	7168.52	-3045.71	-3279.30

Note. Estimated standard errors in parentheses; an extra-multinomial parameter ϕ has been fitted.

TABLE 9.4

Continuation of Table 9.3 Parameter Estimates for Further Two-Level Ordered
Category Models for KS1 Reading Test

	Model CL	Model DL	Model EL
Fixed parameters			
$\theta^{(1)}$	-4.54 (0.12)	-4.90 (0.16)	-4.96 (0.14)
$\theta^{(2)}$	-2.22 (0.09)	-2.20 (0.13)	-2.27 (0.11)
$\theta^{(3)}$	-1.24 (0.09)	-0.97 (0.13)	-1.03 (0.11)
$\theta^{(4)}$	-.04 (0.09)	0.06 (0.13)	-0.01 (0.10)
$\theta^{(5)}$	0.61 (0.09)	0.85 (0.13)	0.79 (0.10)
Baseline tests			
Spelling		0.21 (0.04)	0.20 (0.04)
Reading		0.35 (0.04)	0.36 (0.04)
Writing		0.15 (0.04)	0.17 (0.04)
Number		0.80 (0.04)	0.79 (0.04)
Algebra		0.30 (0.04)	0.30 (0.04)
Shape and space		0.02 (0.04)	0.04 (0.04)
Handling data		0.28 (0.04)	0.29 (0.04)
Gender	-0.61 (0.05)	-0.46 (0.06)	-0.45 (0.06)
Free school meals	-0.71 (0.06)	-0.41 (0.06)	-0.39 (0.06)
Nursery	0.23 (0.06)	-0.08 (0.07)	-0.08 (0.07)
Centred age	0.10 (0.07)	0.01 (0.01)	0.01 (0.01)
EthLang2	-0.01 (0.12)	-0.09 (0.12)	-0.10 (0.12)
EthLang3	0.14 (0.44)	0.15 (0.49)	0.21 (0.48)
EthLang4	0.27 (0.13)	0.22 (0.13)	0.21 (0.13)
EthLang5	-1.10 (0.11)	-0.25 (0.13)	-0.28 (0.12)
EthLang6	0.51 (0.85)	1.12 (0.87)	1.15 (0.86)
EthLang7	-0.44 (0.14)	0.07 (0.15)	0.04 (0.14)
EthLang8	-0.06 (0.27)	-0.08 (0.29)	-0.12 (0.29)
EthLang9	-1.28 (0.17)	-0.29 (0.19)	-0.29 (0.18)
EthLang10	-1.09 (0.41)	-0.26 (0.42)	-0.32 (0.41)
EthLang11	0.55 (0.62)	1.09 (0.67)	1.11 (0.68)
EthLang12	2.02 (1.05)	2.78 (1.04)	2.69 (1.04)
EthLang13	-0.34 (0.39)	0.24 (0.40)	0.22 (0.40)
EthLang14	0.43 (0.65)	0.04 (0.68)	0.02 (0.69)
Baseline aggregate			0.83 (0.09)
Free school meals context			0.38 (0.06)
Random parameters			
School variance: $\hat{\sigma}_u^2$	0.301 (0.052)	1.112 (0.161)	0.522 (0.08)
School % of residual var in lv model	8.4	25.2	13.7
Approx reduction in lev 1 logistic latent var from model AL	75%	69%	68%
Approx rescaled School var using AL model logistic CMS scale	0.216	0.764	0.355
Extra-multinomial $\hat{\phi}$	0.960 (0.009)	0.952 (0.012)	0.943 (0.009)
-2 log-likelihood	-1034.20	-4070.51	-4504.31

9.1 that level 1 response variability has reduced considerably. This is not explicit in Table 9.3 results where the underlying response variable has effectively been rescaled (by restandardizing the level 1 response variance). This impacts on all parameter estimates and in particular the school variance $\hat{\sigma}_u^2$. How much of the change in the latter is due to the scale change and how much is due to a real change consequent to the baseline controls is not apparent (although the change in the school variances of log-odds is real enough).

Changes in the $\hat{\theta}_{(s)}$ as the model develops are instructive in understanding the implicit response scale changes. They account for the proportionality of the odds or on the lv interpretation represent cut points. Suppose the unlikely event that introduction of extra variables and effects into a model did not affect level 1 response variability and hence the scale. Naturally, as usual, there might be changes in existing effect parameters and the school variability. The latter changes would affect the variability in the locations of the conditional level 1 response distributions in model–specified ways. However, the spread within these distributions would be invariant. The relationships between the set of cumulative odds for a given distribution as reflected by the relative sizes of the $\hat{\theta}_{(s)}$ would not be expected to change much. Thus, the latent variable cut point parameters should stay roughly the same, apart from a possible constant additive intercept change due to the introduction of an explanatory variable with a nonzero mean. Thus, we might conclude that apart from such a constant shift, if any changes in the $\hat{\theta}_{(s)}$ were noted, these arise largely as a result of rescaling the lv as a result of a change in the level 1 variance and were otherwise ruled out. Because the introduced variables in BL all have mean zero, a direct comparison may be made between its $\hat{\theta}_{(s)}$ and those of AL. A stretching of the set of values is apparent, indicative of a reduction in level 1 response variability previously noted from Table 9.1. The factors by which the $\hat{\theta}_{(s)}$ are inflated thus provide a rough indication of the extent of rescaling required in the lv as a result of restandardization following a real change in level 1 residual dispersion.

It should be clear that in order to separate out meaningful changes in parameter values from scale changes, some clearer handle on the level 1 change is required. The relationship to cut point parameter changes gives a hint of how this might be done. An approximate method has been devised, full details of which are provided by Fielding and Yang (in press). The key is an attempt to keep the scale of the underlying variable unchanged as the model develops away from the base. From cut point estimates for model AL and the fitted distribution in Table 9.1, we can construct logistic conditional mean scores (CMS) for the categories (Fielding, 1997). The variance of the fitted response distribution using these can then be found (which, due to grouping effects, will usually be a little below the theoretical $\pi^2/3$). For these common CMS scores, and thus using the same scale, the level 1 variance can be calculated from the cut points and a fitted conditional level 1 distribution for further developments of the base model. The reductions in this common scale level 1 variance for the various logit models are given in Table 9.3. Model BL variance is 72% that of AL, implying a level 1 variance reduction of 28%. To account for the rescaling, this factor may be applied to the school estimated variance to yield 0.880. The increase in school variance in the controlled model over the base performance model (0.58) is not uncommon in primary school research. However, it is not as large as may seem from the unscaled estimate. Estimates of the scaled variances for other models are given in Table 9.4. It is also possible to put other parameter estimates on a common footing for comparative purposes by using a square root factor mutiplier. Applied to cut point estimates, it brings them as expected almost into line in most cases.

This discussion has used the example to illustrate points of model interpretation. There is much detailed substantive content in the results pertinent to a debate on educational issues. Only brief comments are made here. Baseline assessments and individual background variables are closely associated so model CL controls for the latter separately. Many ethnic-language dummies represent small groups, are estimated imprecisely, and are not statistically significant. However, there is evidence that fairly large groups of non-English speaking Pakistani and Bangladeshi children are disadvantaged. A net gender gap is evident in favor of girls. Their odds of getting below each grade are about half [exp(-061)] those of boys. Performance is worse for free school meals children and better if they have had nursery education. The rescaled school variance shows a considerable reduction over the base for CL, unlike that for BL. Further, this reduction is greater than for pupil variance. Thus, although the socioeconomic characteristics are closely related to baseline scores, they operate quite differently and thus, cannot be taken as proxy, as is sometimes done. Catchment area characteristics of schools may make them more internally homogenious on these factors than on ability. The importance of recognizing the multilevel structure is thus again emphasized here. Model DL with both sets of factors has a similar level 1 variance and hence, scaling to BL, so parameter estimates are directly comparable. Baseline test coefficients are very similar and it appears that baseline net effects are largely unaffected by extra controls. These extra background variables do have a highly significant effect on progress, even when initial ability has been controlled, (Wald χ^2 is 148.4 on 17 df). Girls also progress more than boys. This is in contrast to an opposite result noted by Fielding (1999) for mathematics KS1. Free school meals is related to progress

TABLE 9.5

"Chances" Distributions: Predicted Percentage Distributions Over KS1
Reading Levels for Some Combinations of Baseline Assessment Levels

KS1 Reading Level	0	1	2C	2B	2A	3
Empirical overall percent	3.6	19.6	17.7	16.6	12.0	30.5
Baselines at level 1						
Number at level 0	8.5	38.9	21.2	15.4	7.7	8.3
Number at level 1	3.2	25.6	21.2	17.6	12.6	19.8
Number at level 2	1.1	12.3	17.8	17.6	13.6	37.4
Baselines at level 2						
Number at level 0	1.1	12.6	18.0	17.6	13.6	36.9
Number at level 1	0.4	5.0	10.1	15.2	13.7	55.5
Number at level 2	0.1	1.8	4.3	8.6	11.1	74.1

net of ability but the same cannot be said of nursery education. Again,
most ethnic-language effects are not statistically discernible. However, the
effects of some of them on attainment levels does not seem to carry over for
adjusted levels of progress. Model EL has two school context effects that
explain a significant extra amount of school variation (rescaled variance
is half that of model DL). Regardless of their individual characteristics,
pupils appear to make more progress when their peer group in school is
more advantaged. A general conclusion of these results is the separate and
differently operating contributions of prior ability and other characteristics,
which the multilevel analysis brings out.   These facts have often been
ignored in policy–related research where, in the absence of prior ability
measures, correlated background factors have been used as proxies.

Two particular analytical uses of these types of model are now
briefly outlined. The Value Added National Project (Fitz-Gibbon, 1997)
recommended using "chances" of achieving levels for students with different
profiles as a useful way of making predictions. This is difficult when points
scores are used. Ordered category models provide a direct way of doing
so because the focus is on the entire distribution over discrete categories.
For illustration, Table 9.5 presents "chances" based on a "value added"
model BL. They might be used when it is desired to adjust for prior
ability measures only. Adjustments for expectations of nonlinear functions
suggested by Hedeker and Gibbons (1994) and by Goldstein (1995, p. 79)
are made before converting the fitted LP for sets of covariate values to
predictions of probabilities. The top of four baseline levels is very rare with
no more than 2.2% in the data for any test. Thus, the illustrations set all

the baseline variables except number at either level 1 or 2. The important influence of number can be seen by allowing it to vary. Such distributions, perhaps with more detailed profiles, when converted to graphical displays, provide a readily understood motivational device. However, they are based on point estimates of model parameters. For any detailed inferences, the uncertainty of estimates should be recognized and confidence bands for the distributions provided. It should also be recognised that a school effect could be included in the profile and will alter the predictions. These fitted "chances" are averaged over schools. Generally, a pupil may be expected to do better or worse than a similarly profiled pupil if a different school is attended.

A second application of the model uses the MULTICAT estimation of the school residuals u_j and their standard errors. The residuals in models with control variables may be used to derive value–added measures although we have a preference for the term "adjusted school effects." The details of the controls to be used and the mechanism by which they operate are an area of vigorous debate in the school effectiveness literature. A model for outcome responses is often likened to an economics production function (Woodhouse & Yang, 2001). The role of a school effect is seen as adding to the raw material characterized by other variables in the model. The model adjusts by controlling for this input quality. Official sources often eschew the use of control using socioeconomic characteristics, which, given the earlier results, may surely be contentious. However, for illustration, Fig. 9.1 shows the estimated school effects (residuals) for model BL using only initial ability as controls. Uncertainty must be recognized in these estimates (Goldstein & Spiegelhalter, 1996). Thus, as suggested by Goldstein and Healy (1995), they are surrounded by the 1.4 standard errors giving overall 95% confidence intervals for sets of such comparisons. These caterpillar diagrams have become very familiar in school research but have, as yet, to make inroads into official publication. They can be routinely implemented in the graphics windows of MlwiN. The scale is that of a logistic variable and could be converted to standard deviation units of outcome on division by $\pi/\sqrt{3}$. Schools and others are used to standardized scores, so fairly easy interpretation of such diagrams should be possible. Scales can also be converted to KS level units if required. The overlap between school bands as also noted in many other published contexts, means that it is difficult to discriminate between the effectiveness of the majority of schools. For screening purposes, there are some obviously extreme schools at either end. These arouse interest and could be subject to further scrutiny. A few of the schools at the top end, for instance, are particularly advantaged in the context effects and socioeconomic variables. These have not been controlled in the graphed results. The diagnosis and analysis of outliers is aided by the

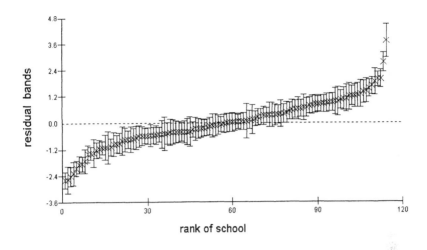

FIG. 9.1. School estimated residual effects for model BL in rank order; Bands are residual +/- 1.4 estimated standard errors

methodological developments of Langford and Lewis (1998) implemented in the graphical interface of MlwiN.

CROSS–CLASSIFIED AND WEIGHTED RANDOM EFFECTS IN MULTILEVEL ORDERED CATEGORY RESPONSE MODELS

Structures, data, and models

At any level in a model, there may be additional hierarchies of units that cut across existing ones and that contribute effects to random variation. The first example to be discussed recognizes the continuity of school effects. The data is available with the MlwiN software. The response, attainment of 3,435 children around 16 years old in 19 secondary schools in Fife Scotland, is graded into 10 ordered categories. The children are also nested in 148 primary schools they previously attended before transfer at around 11 years of age and that may also exert an effect. Thus, children are lodged in a cross-classification of primary and secondary school at level 2. As a point of comparison, the example has been analyzed using continuous response models and grade level scores (1–10) in the MlwiN user guide (Rasbash et al., 1999). Here, ordered category models are used, and in investigation, a preference for the probit link emerged. Thus, the model for cumulative grade probabilities is of the form

$$\Omega^{-1}\left[\gamma_{i(j_1,j_2)(s)}\right] = \theta_s - \left(\sum_{\ell=1}^{L}\beta_\ell x_{\ell i(j_1,j_2)} + u_{j_1} + u_{j_2}\right),$$

$s = 1,2, \ldots,9$. The j_1 and j_2 indices range over secondary and primary schools respectively, which have separate additive random effects at level 2. For inference purposes, these are assumed as usual to be normally and independently distributed with variances $\sigma_{u_1}^{2}$ and $\sigma_{u_2}^{2}$ to be estimated. Level 1 observations, indexed by i, are lodged within cells (j_1, j_2) of the level 2 crossing. More generally, in such models, there may be many ways of crossing, crossings at many levels, and random regression coefficients. The range of complex possibilities was reviewed in Goldstein (1995). Applications with cross-classified effects in continuous response models were discussed in Goldstein and Sammons (1997), Raudenbush (1993), and for binary responses in Yang, Goldstein, and Heath(2000).

The second example to be analyzed uses a subset of data drawn from a study of effectiveness in further education colleges in England and Wales (Belfield, Fielding, & Thomas, 1996). Level one in the structure is entries to subjects at the General Certificate of Education Advanced Level with six ordered grades. The 3,717 entries are nested within 1,522 students and further within 317 subject teaching groups. There is thus a crossing of student and group at level 2. The desirability of disentangling their separate random effects is discussed in Fielding and Yang (in press). There the desirability of unpicking separate teacher effects is also raised. If only one of the 145 teachers taught groups, many of whom taught more than one group, a three-way classification at level 2 would be formed. However, it is the norm that groups are taught by several teachers. Weighted random teacher effects handle this. Logit models of the form

$$\ln\left[\frac{\gamma_{i(j_1,j_2)(s)}}{1 - \gamma_{i(j_1,j_2)(s)}}\right] =$$

$$\theta_s - \left(+\sum_{\ell=1}^{L}\beta_\ell x_{\ell i(j_1,j_2)} + u_{j_1} + u_{j_2} + \sum_{j_3=1}^{J_3} w_{i(j_1,j_2)j_3} u_{j_3}\right),$$

where ($s = 1, 2, \ldots, 5$) are used. The j_1 and j_2 index students and groups. Teacher effects are denoted by u_{j_3} with variance $\sigma_{u_3}^{2}$. Weighted contributions across all teachers are formed, with weights, $w_{i(j_1,j_2)j_3}$, being the proportion of a teaching group timetable taught by teachers. The sum ranges over all J_3 teachers but for each observation, only a few weights will be nonzero.

Model estimation

Theory and methods for recasting cross-classified and weighted random models for continuous response models into hierarchical form have been discussed fully by Raudenbush (1993) and by Goldstein (1987, 1995). A range of commands that can do this and that can be incorporated into macros is available in MlwiN. The *User Guide* (Rasbash et al., 1999) discusses example setups. In the previous examples, the crossings take place in the LP at levels above the observation and thus these reformulations can carry over quite readily into GLMMs for ordered responses. Quasi-likelihood estimation as previously discussed can then be carried out. For technical reasons due to the necessity for imposing prior constraints on variance parameters, the standard ordered response procedures in MULTICAT cannot be directly employed. A stand-alone macro ORDCAT that overcomes this has been written.[1]

Attainment in Scottish secondary schools

For the example results, a value added analysis using only a single explanatory variable is used; STVRQ is a standardized score on a verbal reasoning test taken before entry to secondary school. With the number of categories as large as 10 a linear analysis using points scored responses may be a useful preliminary in model investigations. Two scoring schemes are used in results presented before the probit model in Table 9.6; equal intervals scaled to have mean zero and unit variance on the data and standard normal scores. In all models, polynomial terms in STVRQ were also tried. These and the possibility of a STVRQ coefficient random at the secondary level proved uninteresting. For comparative purposes, in Table 9.7, the probit model has also been rescaled to have level 1 *lv* variance, the same as the normal scores model.

Broad patterns of major significant effects would emerge from a variety of model types even if some assumed features were doubtful (Ezzett & Whitehead, 1991). With as many as 10 categories, it is seen that the scored analyses and the probit model give much the same impression. For more detailed analyses, such as estimation of school effects or predicting grade distributions, the less restrictive probit or other GLMMs would usually be preferred. Broadly, the results confirm other work (Goldstein & Sammons, 1997) on the importance of primary school attended on secondary school performance and progress. In all base performance models

---

[1] ORDCAT and accompanying documentation is available from the author's web page, www.bham.ac.uk/economics/staff/tony.htm., which also has a range of relevant discussion papers. Macros are in a continuous state of development and the latest forthcoming version (Yang et al., 1998) will incoporate the present procedures.

TABLE 9.6

Parameter Estimates in Cross-Classified Models using points scores for
Attainment in 10–Point Categories for 16-Year-Old Children in Secondary
Schools in Fife, Scotland

	Equal Interval Standardized Responses		Standardized Conditional Mean Normal Scores	
	Variance Components	*With Control*	*Variance Components*	*With Control*
Fixed Effects				
Intercept	-0.057 (0.052)	-0.065 (0.022)	-0.054 (0.056)	-0.015 (0.022)
STVRQ		0.696 (0.021)		0.689 (0.012)
Random Effect				
Variance				
Secondary Schools	0.037 (0.014)	0.001 (0.003)	0.035 (0.017)	0.001 (0.002)
Primary Schools	0.120 (0.021)	0.029 (0.006)	0.121 (0.021)	0.032 (0.007)
Pupils	0.867 (0.021)	0.455 (0.011)	0.867 (0.021)	0.460 (0.011)

Note. Estimated standard errors in parentheses.

without STVRQ control, it is seen that primary school variance is over three
times that of secondary schools. When STVRQ, which may be considered a
primary school outcome, is controlled along with other net primary school
effects, the secondary school contribution to secondary school progress
becomes quite small. Goldstein (1995) discussed how such results should
be interpreted with care due to the larger number of primary schools whose
effects are averaged within secondary schools. Nevertheless, the importance
of continuity of educational effects emerges.

The probit and scoring models differ somewhat in detail of parameter
estimates whereas these can be compared. Using the same approximate
method as before but with normal CMS, a *lv* rescaling of the variance
components by 58% may be applied to the probit model with STVRQ
control. It may be noted that this contrasts with the slightly smaller
53% reduction in level 1 variance in both points models. After scaling
the 75% reduction of primary school variance from 0.1555 to 0.0305, after
introduction of STVRQ may be compared with 80% for the equal interval

TABLE 9.7

Parameter Estimates in Cross-Classified Models using Ordered Category Probit Models for Attainment in 10–Point Categories for 16-Year-Old Children in Secondary Schools in Fife, Scotland

	Without STVRQ Control	With STVRQ Control	Rescaled[1] Without Control	With Control
Fixed Effects				
θ_1	-2.074 (0.081)	-2.893 (0.079)	-1.931	-1.962
θ_2	-0.796 (0.066)	-1.240 (0.045)	-0.741	-0.841
θ_3	-0.382 (0.065)	-0.647 (0.042)	-0.376	-0.439
θ_4	-0.188 (0.065)	-0.264 (0.040)	-0.110	-0.179
θ_5	0.120 (0.065)	0.087 (0.039)	0.112	0.059
θ_6	0.313 (0.065)	0.374 (0.040)	0.291	0.254
θ_7	0.505 (0.065)	0.658 (0.041)	0.470	0.446
θ_8	0.715 (0.066)	0.965 (0.043)	0.660	0.665
θ_9	1.021 (0.067)	1.396 (0.046)	0.950	0.945
STVRQ		1.112 (0.029)		0.745
Random Effect Variance				
Secondary Schools	0.045 (0.021)	0.003 (0.006)	0.039	0.001
Primary Schools	0.156 (0.027)	0.053 (0.016)	0.135	0.024
After Rescaling by x = 58%				
Secondary Schools		0.002		
Primary Schools		0.031		
Extra-multinomial ϕ	0.970 (0.008)	1.174 (0.012)		

Note. Estimated standard errors in parentheses.

[1] The rescaling is to have the same level 1 lv variance as variance estimated in parallel Normal Scores points model in Table 9.6

model and 73% for the normal scores model. The rescaling of the probit results to the level 1 variances of the normal points model on the right–hand side of Table 9.7 shows that there may be a considerable difference in the estimated marginal net impact of initial ability on the response. The probit model estimate of 1.212 for the STVRQ parameter is approximated to 0.75 on a comparable scale to the smaller estimates of 0.69 for the standard points models. This would make a substantial difference to chance predictions that could be carried out as for the previous examples.

Group student and teacher effects on GCE advanced level

The first two columns of Table 9.8 present estimates for a base and elaborated hierarchical logit model of entries within teaching groups. Although useful as a point of comparison, they ignore lack of independence of the entries across groups due to shared and unmeasured student effects. Also, they do not attempt to disentangle group and teacher effects. The results for comparable models, which incorporate crossed student effects and weighted teacher effects, are given in the second two columns.

The same set of explanatory variables and effects is used in each case, although, as usual, they ensue from fairly deep model investigation using available data. Details of the variables are fairly explicit in the legend and body of Table 9.8. Subjects are dummy variables related to a Social Science base. The base for Instition dummies is a medium sized Further Education College. FEC, TC, SFC, denotes Further Education, Tertiary and Sixth Form Colleges. Although the main objective of this example is to discuss the elaboration of random effects, some brief comment may be made on important fixed effects. The main effect is a measure of students' prior attainment STGC, a standardized average of grade points on a number of General Certificate of Secondary Education subjects taken by students usually just before they embark on the Advanced Levels. A quadratic term is also required for this, which may reflect the marked skewness of STGC to a ceiling. It may be noted that girls seem to have worse net performance and hence progress when prior ability and other factors are controlled. By contrast, although not illustrated here, girls have higher unadjusted raw performance. Also, the significant negative interaction between STGC and gender indicate that lower ability girls make more progress than similar boys but vice versa at higher ability levels. The results indicate some subject group differences as found in other research but these are not elaborated here. The six colleges in the data represented by dummy fixed effects in the model cover a range of types and sizes. Some net differences between the colleges are evident, but there are too few to facilitate generalizations. They have been included in the model as relevant block adjustment controls.

The change in the student variable fixed parameter estimates after student and teacher effects are introduced is in line with change of cut point estimates and seems indicative of rescaling only. However, there are some uneven changes in subject and college dummies, which are not consonant with scale changes alone. Many explanations not fully explored here may be suggested and are capable of being fully investigated. Clustering of certain types of student in certain subjects or colleges may have been confounded before control of unmeasured student effects was exercised. Types of teacher may also be inextricably bound up with these factors that are specifically separated when there are controls for teacher effects. There is much complexity of interest here that could possibly be unraveled by much deeper investigation of the nature of the educational process. The changes in standard error estimates with more detailed variance specification, although not large, may also be noted.

The variance component estimates across the four models in Table 9.8 raise relevant issues. In the teaching group only models A and B, a lv scaling reveals that the covariates reduce entry variance by 30% from that of the base model The reduction in group variance will be similar as is seen by the similar percentages of residual variance attributable to groups (17.7 and 18.2) before and after covariate control. In Model C with the introduction of student and teacher effects into the base model, the relative contribution of groups is much reduced. Some of this may be taken up by the common influence of student effects. Residual entry variation is still 79% of the total, much of which may be explained by considerable variation of grades within students, and that detail is often lost when aggregate student performances are the focus of enquiry. The teacher variation separate from group variation and relative to it is seen to be considerable. It must be noted though that the teacher variance contribution to an entry observation is not conventionally additive. Denoting the variance parameter estimate as $\hat{\sigma}_T{}^2$, the contribution will be

$$\hat{\sigma}_T^2 \sum_{j_3=1}^{J_3} w^2_{i(j_1, j_2)j_3}.$$

For instance, in Model D introducing the covariates, where $\hat{\sigma}_T{}^2 = 0.4521$, a group with three equally weighted teachers will contribute 0.151. However, this is still larger than the group variance of 0.128. Introduction of covariates impacts more on reducing group and student variance although lv rescaling results are not displayed here. The teacher variation is now relatively more important (11% vs. 8.4%). There is much more detail that could be further investigated. However, the model's ability to isolate teacher effects suggests the broad and important conclusion that teachers do matter. The methodological tools available form an opportunity to investigate this if more data was available. Some conventional teacher characteristics, such as age, gender, training, educational levels, and length of service are available in the data but do not on investigation appear to explain much teacher variation.

The role of extramultinomial variation

An extramultinomial parameter has been used but so far barely commented on. Simulations with multinomial structures have indicated that estimation of all model parameters may be improved, if it is left unconstrained (Fielding & Yang, in press; Yang, 1997). Wright (1997) also showed that for sparse, unbalanced, but multinomial structures, an estimate a bit different from unity often emerges. For these reasons we recommend it usually be let free. The provenance of extravariation is often discussed in terms of misspecified probabilities in controlled experimental designs (e.g., Williams, 1982). For models of complex multilevel survey data that can only ever be approximations, the issue is not so clear cut and misspecification can occur in a variety of ways that are complicated to unravel. This is beyond the present chapter. Most estimates in the results are not too different from unity.

CONCLUDING REMARKS

The examples discussed have shown the potentiality of ordinal GLMMs for fairly complex structures. It is known that the quasi-likelihood procedures used can be biased but Yang (1997) has shown that PQL2 estimation is satisfactory for many situations. Bias reduction bootstrapping procedures according to Kuk (1995) may be used although they are often computationally intensive and thus practically infeasible. They have been applied to the PQL2 results of Table 9.3 where it is found that bias is minimal.

As research develops, there is clearly room for improved methods of estimation. A promising approach adapts PQL2 for cross-classified models by data augmentation and has been applied to binary responses by Clayton and Rasbash (1999). This also incorporates the Bayesian approaches of Monte Carlo Markov Chain (MCMC) estimation. An approach under current investigation that promises unbiased estimation casts the ordered category models into a standard linear form for the lv and simulates this variable at each stage of a standard multilevel estimation. The current estimation procedures available in MlwiN seem satisfactory for the moment but these sorts of ongoing methodological investigation will add to the

TABLE 9.8

Parameter Estimates for Cross-Classified and Weighted Random Effects Models for Performance in Subjects at General Certificate of Education at Advanced Level in Six Colleges for Postcompulsory School–Aged Students

	Model A	Model B	Model C	Model D
Fixed effects				
θ_1	-1.67 (0.07)	-1.54 (0.28)	-1.66 (0.08)	-1.57 (0.28)
θ_2	-0.73 (0.06)	-0.47 (0.28)	-0.72 (0.08)	-0.51 (0.28)
θ_3	0.17 (0.06)	0.59 (0.28)	0.19 (0.08)	0.53 (0.28)
θ_4	1.08 (0.06)	1.67 (0.28)	1.09 (0.08)	1.58 (0.28)
θ_5	2.48 (0.08)	3.32 (0.29)	2.48 (0.09)	3.20 (0.29)
STGC		1.33 (0.05)		1.32 (0.05)
STGC squared		0.27 (0.02)		0.27 (0.02)
Female gender		-0.12 (0.05)		-0.14 (0.07)
Interaction of STGC and gender		-0.18 (0.06)		-0.20 (0.06)
SUBJECTS:				
Art, Design & Tech.		-0.08 (0.20)		-0.05 (0.21)
Mathematics		-0.40 (0.17)		-0.17 (0.24)
Sciences		-0.38 (0.16)		-0.41 (0.18)
Humanities		0.12 (0.16)		0.13 (0.18)
Languages		-0.49 (0.23)		-0.27 (0.26)
General studies		-0.52 (0.40)		-0.44 (0.38)
COLLEGES:				
Large FEC		0.16 (0.29)		0.34 (0.39)
Medium–sized TC		0.99 (0.30)		0.89 (0.31)
Small SFC		0.85 (0.31)		0.70 (0.34)
Medium–sized SFC		-0.12 (0.29)		-0.59 (0.34)
Large SFC		0.58 (0.26)		0.33 (0.29)
Random Effects				
Variance				
Teaching groups	0.708(0.079)	0.731(0.081)	0.215(0.071)	0.128(0.061)
% of lv residual variance	17.7	18.2	5.2	3.1
Students			0.279(0.116)	0.241(0.113)
% of lv residual variance			6.7	5.9
Teachers			0.349(0.162)	0.452(0.158)
% of lv residual variance			8.4	11.0
Extra-multinomial	0.953 (0.010)	0.955 (0.010)	0.955 (0.010)	0.971 (0.010)

Note. Estimated standard errors in parentheses.

Note. The response is the 6 pt graded result of a subject entry. There are 3717 entries within 317 subject teaching groups from 1522 students and 145 teachers. The base for Subject Group dummies is Social Sciences. The base for Institution dummies is medium sized Further Education College: FEC, TC, SFC denotes Further Education, Tertiary and Sixth Form Colleges.

armory of tools for complex model structures.

ACKNOWLEDGMENTS

Part of this work was done during a Visiting Research Fellowship at the Multilevel Models Project, University of London Institute of Education supported by the UK Economic ad Social Research Council under award H51944500497. Thanks are due to Harvey Goldstein, Min Yang, and Jon Rasbash for their constant advice and stimulus.

REFERENCES

Aitkin, M. (1999). A general maximum likelihood analysis of variance components in generalized linear models. *Biometrics, 55*, 117–128.

Anderson, D. A., & Aitkin, M. (1985). Variance component models with binary responses *Journal of the Royal Statistical Society, 47*(Series B), 203–210.

Belfield, C., Fielding, A., & Thomas, H. (1996). *Costs and performance of A level provision in colleges: Research report for the Association of Principals in Sixth Form Colleges.* School of Education, University of Birmingham.

Bryk, A. S., & Raudenbush, S. W. (1992). *Hierarchical linear models.* Newbury Park, CA: Sage.

Clayton, D. J., & Rasbash, J. (1999). Estimation in large crossed random effects models by data augmentation. *Journal of the Royal Statistical Society, 162* (3, Series A), 425–436.

Ezzett, F., & Whitehead, J. (1991). A random effects model for ordinal responses from a crossover trial. *Statistics in Medicine, 10*, 901–907.

Fielding, A. (1997). On scoring ordered classifications. *British Journal of Mathematical and Statistical Psychology, 50*, 285–307.

Fielding, A. (1999). Why use arbitrary points scores?: Ordered categories in models of educational progress. *Journal of the Royal Statistical Society, 162* (3, Series A), 303–328.

Fielding, A. (2000). *Scored and ordinal category resposes in random effects models* (Discussion Paper 00-02). University of Birmingham, United Kingdom. Department of Economics.

Fielding, A., & Yang, M. (in press). *Random effects models for ordered category responses and complex structures in educational progress.*

Fitz-Gibbon, C. T. (1997). *The value added national project: Final report.* London: Schools Curriculum and Assessment Authority.

Goldstein, H. (1987). Multilevel covariance component models. *Biometrika, 74,* 430–431.

Goldstein, H. (1995). *Multilevel Statistical Models* (2nd ed.). London: Arnold.

Goldstein, H., & Healy, M. J. R. (1995). The graphical presentation of a collection of means. *Journal of the Royal Statistical Society, 15* (8, Series A), 175–177.

Goldstein, H., & Sammons, P. (1997). The influence of secondary and junior schools on sixteen year examination performance: A cross-classified multilevel analysis. *School Effectiveness and School Improvement, 8* (2), 219–230.

Goldstein, H., & Spiegelhalter, D. J. (1996). League tables and their limitations: Statistical issues in comparisons of institutional performance (with discussion). *Journal of the Royal Statistical Society, 159* (3, Series A), 85–443.

Harville, D. A., & Mee, R. W. (1984). A mixed model procedure for analysing ordered categorical data. *Biometrics, 40,* 393–408.

Hedeker, D., & Gibbons, R. D. (1994). A random-effects ordinal regression model for multilevel analysis. *Biometrics, 50,* 933–944.

Hedeker, D., & Gibbons, R. D. (1996). MIXOR: A computer program for mixed effects ordinal regression analysis. *Computer Methods and Programs in Biomedicine, 49,* 157–176.

Hedeker, D., & Mermelstein, R. J. (1998). A multilevel thresholds of change model for analysis of stages of Change data. *Multivariate Behavioral Research, 33* (4), 427–455.

Jansen, J. (1990). On the statistical analysis of ordinal data when extravariation is present. *Applied Statistics, 39*(1), 75–84.

Kuk, A. Y. C. (1995). Asymptotically unbiased estimation in generalized linear models with random effects. *Journal of the Royal Statistical Society, 57* (Series B), 395–407. Langford, I. H., & Lewis, T. (1998). Outliers in multilevel data (with discussion). *Journal of the Royal Statistical Society, 161* (Series A), 121–160.

McCullagh, P., & Nelder, J. A. (1989). Generalized linear models (2nd ed.). London: Chapman & Hall.

Rasbash, J., Browne, W., Goldstein, H., Yang, M., Plewis, I., Healy, M., Woodhouse, G., & Draper, D. (1999). *A user's guide to MlwiN* (Version 2.0). Multilevel Models Project, Institute of Education, University of London.

Raudenbush, S. W. (1993). A crossed random effects model for unbalanced data with applications in cross-sectional and logitudinal research.

Journal of Educational Statistics, 18 (4), 321–349.

Stewart, M. B. (1983). On least squares estimation when the dependent variable is grouped. *Review of Economic Studies, 50,* 737–753.

Willams, D. A. (1982). Extra-binomial variation in logistic linear models. *Applied Statistics, 31,* 144–148.

Woodhouse, G., & Yang, M. (2001). Progress from GCSE to A and AS level: institutional and gender differences, and trends over time. *British Journal of Educational Research, 27* (3), 245–267..

Wright, D. (1997). Extra-binomial variation in multilevel logistic models with sparse structures. *British Journal of Mathematical and Statistical Psychology, 50,* 21–29.

Yang, M. (1997). Multilevel models for multiple category responses by MLN. *Multilevel Models Newsletter, 9* (1), 9–15.

Yang, M. (2000). Multilevel models for multiple response categories. In H. Goldstein & A. Leyland (Eds.), *Multilevel modelling in the health sciences* (pp. 107–123). London: Arnold.

Yang, M., Goldstein, H., & Heath, A. (2000) Multilevel models for repeated binary outcomes: attitudes and voting over the electoral cycle. *Journal of the Royal Statistical Society, 163* (1, Series A), 49–62.

Yang, M., Fielding, A., & Goldstein, H. (in press-a). *Multilevel ordinal models for educational data.*

Yang, M., Rasbash, J., Goldstein, H., & Fielding A. (1998). *MlwiN macros for advanced multilevel modelling* (Version 2.0).

10

Bootstrapping the Effect of Measurement Errors on Apparent Aggregated Group-Level Effects

Dougal Hutchison

National Foundation for Educational Research in England and Wales

In much social research, the observations are treated as if they were made without error. Yet, it is widely accepted that this assumption does not hold universally (Woodhouse, Yang, Goldstein, Rasbash, & Pan, 1996). It has been shown that taking account of measurement error in the analysis of educational effects can change and indeed reverse conclusions (Goldstein, 1979). However, such studies are based on single-level regression models. Most educational research data has a hierarchical structure and, is most appropriately analyzed by multilevel models.

A two-level linear model for the true or "latent" values x_{ij} and y_{ij} is given by

$$y_{ij} = x_{ij}\beta + w_{ij} \qquad (10.1)$$

where w_{ij} is the sum of two uncorrelated random variables, one for each level, each assumed to have zero mean and constant variance in a variance components model.

The true or latent values x_{ij} and y_{ij} in Equation 10.1 are observed with measurement error giving observed values X_{ij} and Y_{ij} where

$$X_{ij} = x_{ij} + m_{ij};$$

$$Y_{ij} = y_{ij} + q_{ij} = x_{ij}\beta + e_{ij}; \quad e_{ij} = w_{ij} + q_{ij}; \quad E(q) = E(m) = 0 \quad (10.2)$$

and m_{ij} and q_{ij} are measurement errors, uncorrelated with the true or latent value. If one attempts to estimate the parameters β of this equation on observed data, then one is not in fact estimating Equation 10.1, but a different equation

$$Y_{ij} = X_{ij}\gamma + E_{ij} \tag{10.3}$$

where γ is the OLS regression coefficient of Y_{ij} on X_{ij}, and E_{ij} is the observed residual, and the resultant estimates γ are biased for β (see, e.g., Woodhouse et al., 1996).

Woodhouse et al. (1996) provided a corrected estimate for $\hat{\beta}$ when error variances and covariances are known. Their formulae are described in the following section.

MEASUREMENT ERROR IN LEVEL–TWO AGGREGATED VARIABLES

Where there is a level-2 variable that is the mean of a level-1 variable,

$$X_{2j} = n_j^{-1} \sum_i X_{1ij},$$

and the estimated level-1 reliability [1] of X_{1ij} is given by

$$\rho_1(X_1) = R_1,$$

then the measurement error variance of the aggregated variable X_{2j} is given by

$$\sigma_1^2(m_2) = \frac{\sigma_w^2}{n_j}\left(1 - R_1\frac{n_j - 1}{N_j - 1}\right) \tag{10.4}$$

where $\sigma_{(w)}^2$ is the level-1 variation in X_{1ij}, and $\sigma_1^2(m)$ is the level-1 measurement error variation in X_{1ij} (Woodhouse et al., 1996). The covariance between level-1 and level-2 errors is given by

$$Cov(m_1, m_2) = n_j^{-1}(1 - R_1)\sigma_w^2(X_1). \tag{10.5}$$

---

[1]Level-1 reliability coefficients are quoted because the development in Woodhouse et al. (1996) is in these terms. Level-1 coefficients are not quite the same as the usual reliability figures quoted.

These formulae for level-2 measurement error are used in the results described here.

Application of measurement error models in the present study

This chapter investigates the effect of measurement error in the continuous variables in the model. Space does not permit the description of measurement error (misclassification) in categorical variables. This is considered elsewhere (Hutchison, 1998). The original intention was to look also at the effect of measurement error in the dependent variable, but preliminary analyses showed that, for the models considered here, and a fairly wide range of variation, adjusting for measurement error in the dependent variable had minimal effect on the fixed coefficients. For this reason, this question is not considered any further.

Analyses were carried out using MLn (Rasbash & Woodhouse, 1995) to look at the effect of adjusting for measurement error on the models. The MLn program does not presently provide likelihood ratio estimates for the errors-in-variables case, so the standard errors provided by the program were used instead. To verify the program standard errors for coefficients, bootstrap analyses were also carried out. A brief description of the bootstrap and how the bootstrap may be combined with errors-in-variables models, is now given.

Bootstrapping and errors-in-variables models

As far as the author is aware there are currently (January, 2000) no published articles combining bootstrapping, measurement error, and multilevel models. Examples combining two of these are Haukka (1995; errors and bootstrap), Waclawiw and Liang (1994), Carpenter, Goldstein, and Rabash (1999), Meijer, van der Leeden, and Busing, (1995; multilevel and bootstrap), Woodhouse et al., (1996), and Goldstein (1995; multilevel and errors in variables).

Some discussion about how bootstrapping may be achieved in a model that is both multilevel and has errors in variables is therefore in order. Emphasising first the errors-in-variables aspect, the model can be written, as in Equations 10.1 and 10.2 above

$$X_{ij} = x_{ij} + m_{ij}$$

$$Y_{ij} = y_{ij} + q_{ij} = x_{ij}\beta + e_{ij}; \quad e_{ij} = w_{ij} + q_{ij}; \quad E(q) = E(m) = 0 \quad (10.6)$$

Both the bootstrap and errors-in-variables approaches depend on the conceptualization of random variables. Before applying bootstrap techniques to errors-in-variables models, we should investigate the following: (a) The extent to which the different conceptualizations of random elements in the two approaches are compatible, and (b) the extent to which both bootstrap and errors-in-variables procedures can be applied at the same time to the multilevel model.

RANDOM ELEMENTS IN BOOTSTRAP AND ERRORS-IN-VARIABLES

Random elements in the bootstrap

Applied to regression, bootstrapping can arise under two main different paradigms, depending on assumptions about the regressor variables (Mooney & Duval, 1993)

In the classic case, the resampling residuals model, the regressor variables are treated as fixed (see e.g., Draper & Smith, 1981), and the resampling applies to the stochastic portion of the model, that is, the residuals comparing the actual and fitted data. There are two subcases of the way in which this is implemented, namely by direct resampling of the residuals, and by sampling from the parametric distribution assumed by the model. This latter is sometimes referred to as the parametric bootstrap.

Mooney and Duval (1993) argued that in much social science analysis, resampling units, Y, X may be the most appropriate. x, m, and e are in effect resampled with each unit, and it is assumed that the sampling distribution of m and e are representative of their modeling distributions. This is sometimes described as the nonparametric bootstrap. As discussed later, there can be a variety of models under the general description of nonparametric.

Random elements in the errors-in-variables model

Fuller (1987) described two main approaches, the structural and functional models, depending on the assumptions made about the latent variables.
1. In the functional approach, it is assumed that the latent variables x are fixed.
2. Conversely in the structural model, the x and m are assumed to be independent drawings from $N\{0, \sigma_x{}^2\}, N\{0, \sigma_m{}^2\}$ distributions. It is also assumed that the $\mathbf{e} = \mathbf{w} + \mathbf{q}$ vector is independent of the vector $\mathbf{x}$. Since m and x and e are considered as drawings from normal distributions, they also follow normal distributions.

Assumptions 1 and 2 here parallel the two bootstrap cases, except that here, the latent variables x are considered, rather than the observed variables X as before.

Note that the parallel is not exact for units resampling. In an actual structural model, the three elements x, u, and e are in effect sampled independently, so that sampling the same unit twice would give the same x, but different values of u and e. In a bootstrap resampling, picking the same unit twice gives the same value of u and e as well as of x.

IMPLEMENTING THE BOOTSTRAP ON ERRORS-IN-VARIABLES IN A MULTILEVEL MODEL

The next stage is to determine how bootstrapping procedures can be applied to the problem in hand, and a number of possible approaches are discussed later. The resampling residuals model is treated first, and then the resampling units.

Resampling residuals

This approach to bootstrapping regression estimates values $\hat{\beta}$ of β and the residuals r_{ij} (or their distribution) from the original data. Given an estimate $\hat{\underline{x}}_{ij}$ of $\underline{x}_{ij}$ (described later) for each case in each resample, it then adds resampled residuals $r_{ij}{}^*$ from the original regression to the fitted $\hat{Y}_{ij} = E\left(Y_{ij}|\hat{\underline{x}}_{ij}\right)$ to give $Y_{ij}{}^*$, where $Y_{ij}{}^* = \hat{Y}_{ij} + r_{ij}{}^* = \hat{\underline{x}}_{ij}\underline{\hat{\beta}} + r_{ij}{}^*$. The corresponding value of β^* is then estimated by regression of $Y_{ij}{}^*$ on $\hat{x}_{ij}$. In addition, when treating errors-in-variables models, random terms from the $\underline{X}$ error distributions are added to the estimated independent variables after generating $Y_{ij}{}^*$ but before running the regression. In the development here, only a single independent variable measured with error is used.

The estimated value X of x in the bootstrapping makes little difference in practice (Hutchison, 1998), so a simple assumption, setting $\underline{\hat{X}} = X$, the observed value, was used.

Parametric Bootstrap. The bootstrapping is carried out as follows. For given proposed values of x_{ij}, x_j, namely $\hat{x}_{ij}$, $\hat{x}_j$, Y^* is then simulated given the estimated $Y_{ij}{}^* = \hat{x}_{ij}\beta + r_p{}^*$ where $r_p{}^*$ is a sum of two random error terms, one for each level. This is carried out using the command SIMU in Mln. A term m_{1ij} is selected from the posited distribution of $N(0, \sigma_{m_1}^2)$ and is added to $\hat{x}_{ij}$, to give $X_{ij}{}^*$. Similarly, a term m_{2j} is selected from $N(0, \sigma_{m_2}^2)$ and added to give $X_j{}^*$ if this term is included in the equation. Then $\underline{\beta^*}$ is estimated in the usual way by multilevel regression of Y^* on the independent variables $\underline{X^*}$ correcting for measurement error. This process is replicated a number of times to give a distribution for $\underline{\beta^*}$.

Conditional Semiparametric Model (CSP). An alternative to sampling residuals from a hypothesised distribution (for example, the normal) is to resample from estimated residuals in the analysis. Starting again with the one-level model, the observed residuals are estimated as

$$
\begin{aligned}
Y_i - \hat{Y}_i &= Y_i - X_i\hat{\beta} \\
&= Y_i - x_i\hat{\beta} - m_i\hat{\beta}.
\end{aligned} \tag{10.7}
$$

The variance of these observed residuals is given by $V\,(obs) \approx \sigma^2 + \beta^2\sigma_m^2$, treating $\hat{\beta}$ as fixed $= \beta$ as is usual in this approach (Mooney & Duval, 1993).

To get unbiased residuals, one can deflate the observed residuals by a factor

$$
\left(\frac{\sigma^2}{\sigma^2 + \beta^2\sigma_m^2}\right)^{\frac{1}{2}}.
$$

The situation for two-level residuals is more complicated. In outline, one subtracts predicted from actual values of Y for each case to give the observed residual, and estimates level-1 and level-2 residuals from this. The intuitive procedure, to take level-2 predicted residuals as the mean of the observed raw residuals for each level-2 unit, and the level-1 residuals as the difference between the observed raw and level-2 residuals, would be expected to give biased results for level-1 and level-2 variances because of variance migration. Variance migration (Hutchison, 1998) occurs when the observed variation at one level contains elements from another level. Preliminary analyses (not shown here) on the no errors-in-variables case confirmed this.

The observed residuals were modified to allow for this bias. The first stage was to estimate the variance of the observed residuals. A simple two-level model, with two predictor variables, x_{ij}, μ_j, where μ_j is the true mean of x_{ij} in the jth level-2 unit, is given by

$$
Y_{ij} = x_{ij}\beta_1 + \mu_j\beta_2 + u_i + e_{ij} \tag{10.8}
$$

where x_{ij} is measured with error by X_{ij} such that $X_{ij} = x_{ij} + m_{ij}$, and μ_j is measured by $X_{j.}$ where $X_{j.} = n_j^{-1}\sum_i X_{ij}$ where the sum is taken over the observed cases in the ith level-2 unit.

$$
\hat{r}_{ij} \quad \text{were estimated as} \quad \hat{r}_{ij} = Y_{ij} - \hat{Y}_{ij}. \tag{10.9}
$$

Level-2 observed residuals $\hat{r}_j$ are given by

$$\hat{r}_j = n_j^{-1} \sum_i \hat{r}_{ij}, \tag{10.10}$$

and level-1 observed residuals $\tilde{r}_{ij}$ are given by

$$\tilde{r}_{ij} = \hat{r}_{ij} - \hat{r}_j. \tag{10.11}$$

$$V(L1) = \frac{n_j - 1}{n_j}\sigma^2 + \beta_1^2 \frac{n_j - 1}{n_j}\sigma_m^2$$

$$
\begin{aligned}
V(L2) \quad = \quad & (\beta_1 + \beta_2)^2\, n_j^{-1}\,(1 - \rho_1)\,\sigma_w^2\,(X) \\
+ \quad & \beta_2^2 \frac{N_j - n_j}{N_j - 1} n_j^{-1}\rho_1\sigma_w^2\,(X) \\
+ \quad & \sigma_u^2 + n_j^{-1}\sigma^2 \tag{10.12}
\end{aligned}
$$

where $\sigma_w^2(X)$ is the within level-2 units variance of X, and ρ_1 is the within level-2 units reliability of X (Woodhouse et al., 1996).

The variances of level-2, and to a lesser extent of level 1, and the observed residuals differ between level-2 units. They are made equal by deflating by the factors $[\hat{\sigma}^2/V(L1)]^{\frac{1}{2}}$ and $[\hat{\sigma}_u^2/V(L2)]^{\frac{1}{2}}$ to give shrunken level-1 and level-2 residuals $\hat{e}_{ij}$ and $\hat{u}_j$ respectively, where $\hat{u}_j = \left(\frac{\hat{\sigma}_u^2}{\hat{V}(L2)}\right)^{\frac{1}{2}} \hat{r}_j$, and $\hat{e}_{ij} = \left(\frac{\sigma^2}{V(L1)}\right)^{\frac{1}{2}}$.

Because the $\hat{u}_j$ are in the nature of regression predictions, they do not have the appropriate variance, so their variance was increased by adding a random term v, $v \sim N(0, \sigma_v^2)$, where $\sigma_v^2 = Max(0, \sigma_u^2 - V(\hat{u}_i))$, where σ_u^2 is the level-2 variance estimated by the program, so that $\hat{u}_j^* = \hat{u}_j + v_j$. Another possibility (Carpenter et al., 1999) would be to inflate the partially adjusted l-2 residuals by the ratio $\frac{V(\hat{u}_j)}{\hat{\sigma}_u^2}$.

Similarly, level-1 augmented residuals were estimated as $e_{ij}^* = \hat{e}_{ij} + o_{ij}$, where $\hat{e}_{ij}$ is the partially adjusted residual, and $o \sim N(0, \sigma_o^2)$ where $\sigma_o^2 = Max[0, \hat{\sigma}^2 - V(\hat{e}_{ij})]$, where $\hat{\sigma}^2$ is estimated from the program, to give the appropriate level-1 variance.

Resampling units

Simple Random Resampling (SRR)

The standard nonparametric bootstrap involves resampling the individual population elements using simple random sampling with replacement and estimating the beta coefficients for each resample (Mooney & Duval, 1993).

The simple random resample approach is suspect from a statististical perspective. A simple random sample from an infinite population does not reproduce the population structure, in particular the arrangement of pupils within schools. If there are an infinite number of schools, then a simple random sample will give a zero possibility of two pupils being drawn from the same school, whereas the multilevel model generally has substantial grouping of pupils within schools. Hutchison (1998) showed that this approach again gives biased estimates because of variance migration. A different nonparametric resampling approach, which takes account of the hierarchical structure, is described in the next section.

Two-stage random resampling (TSRR) M schools are resampled with equal probability with replacement from the population of M schools in the study. Each school hit is treated separately. Within each selected school j, n_j pupils are resampled. This gives equal probability of selection for each pupil in the population, and preserves the school size for each selected school, although it does give a variable number of pupils in each resample, because some samples will pick more larger schools, and some samples more smaller ones.

Two possible methods allowing for variance migration were considered:

1. Mean group scores were defined as the original mean of the pupils in that school and individual scores were defined as the sum of the original group mean, and the deviation of the selected score from the selected group mean. Then level-2 variance is equal to that of the population $\sigma_u{}^2 + n_j^{-1}\sigma^2$ and the estimate $\frac{n_j}{n_j-1}\hat{V}(e)$ is unbiased for σ^2 in the jth group.

2. Rather than adjust the data collection step, an alternative is to adjust the measurement error correction procedure. A two-stage resample is taken, resampling with replacement first schools and pupils within them. $\hat{\beta}$ is estimated from this. This method is the one used in this chapter, and is now described. Level 1 is considered first.

Level 1. For measurement error, as in Equation 10.1, the X value in each resampled case consists of a true component and a measurement error component,

$$X_{ij} = x_{ij} + m_{ij},$$

so that picking a case involves picking a pair consisting of true score and measurement error. Provided that there is no level-2 component in m_{ij}, then the error realizations can be treated as exchangeable, and in the cases picked, can be expected to have a distribution of an SRS resampling. So

$$\sigma_m{}^2 = (1 - \rho) V(X)$$

$$= (1 - \rho_1) V_w (X)$$

where ρ_1 is the within level-1 reliability.

For residuals, the bootstrapping procedure involves resampling from the observed within-group residuals. Their expected variance for the jth level-2 unit is

$$\frac{n_j - 1}{n_j} \sigma^2 \quad ,$$

so resampling with replacement from these gives an observed level-1 variance of

$$\frac{n_j - 1}{n_j} \left(\frac{n_j - 1}{n_j} \sigma^2 \right) \quad .$$

The observed level-1 residual from the jth group is thus effectively resampled from a sample with variance

$$\frac{n_j - 1}{n_j} \sigma^2 .$$

This bias is corrected by adjusting the value of the level-1 variance estimated by the bootstrap procedure by multiplying by $1/\left(\frac{n_j - 1}{n_j} \right)$. Level 2. For measurement error, if X_{tj} is the observed resampled mean of the jth level-2 unit in the TSRR procedure, and X_j is the mean of the jth l-2 unit in the original sample, then

$$X_{tj} - \mu_j = (X_{tj} - X_j) + \left(X_j - \mu_j \right)$$
$$= \quad T_2 \quad + \quad T_1$$

where μ_j is the true mean value of x_{ij}, the true value corresponding to the observed X_{ij}. T_1 is the original measurement error. T_2 is the additional measurement error arising from the bootstrapping procedure.

$$V (T_2) \quad = V (X_{tj} - X_j)$$
$$= n_j^{-1} \frac{n_j - 1}{n_j} V_w (X_{ij}) .$$

Similarly, there is also an additional level-1–level-2 covariance term.

$$n_j^{-1} \frac{n_j - 1}{n_j} \sigma_m^{\ 2}$$

For residuals, the observed level-2 residual variance, in addition to the original l-2 variance, also contains an extra term

$$n_j^{-1} \frac{n_j - 1}{n_j} \sigma^2 .$$

The correction terms contain elements such as the true residual variance, which are the eventual outcome of the procedures. In mathematical terms $\hat{\beta}^+ = f(Y, X, E)$ where $\beta^+ = (\beta, \sigma_u^2, \sigma^2)$ is the multilevel model solution, allowing for measurement error E, and $E = g(X, \beta^+)$ is a function of X and β^+.

An iterative procedure was used to provide a solution for β^+. First, the estimated variances were corrected using the uncorrected estimates, and then the process was repeated using these updated estimates until it converged. Thus, $\beta_0^+ = f(X, 0)$ where 0 is the zero matrix.

$$E_{i+1} = g(X, \beta_i^+)$$
$$\hat{\beta}_{i+1}^+ = f(X, \beta_i^+)$$

There was little difference between the first and second stage estimates, so only one iteration was used.

Nonparametric maximum likelihood model (NPMLE)

We attempted to devise a method to deal directly with two-level measurement error within a nonparametric framework. In the event this did not prove possible because random parts of the model arise in two different ways; (a) by sampling without replacement from a finite population (l-2 Sampling error) or (b) by taking a realization of a random quantity, that is, effectively with replacement sampling (l-1 measurement error and model error term).

These types of random elements have to be generated in different ways, and unit resampling methods which treat all elements equally do not do this. One could alter the details of the error-correction procedure to cope with this, but the conceptual simplicity of the method would be lost.

Level-2 units resampling

The final units resampling method involves resampling entire level-2 units intact. For each replication, j level-2 units were selected with replacement from the existing sample with equal probability. This is described as level-2 only resampling.

Bias-correction methods

A general approach is to use the bias-correction method based on that of Kuk (1995). The procedure was as follows.

Stage 1. A first estimate of β, β_0, was obtained from the program MLn.

Stage 2. A realization Y_{ijm}^* was obtained using SIMU in MLn and β_0.

Stage 3. Random error was added to the independent score variables using the formulae in Woodhouse et al. (1996), quoted earlier.

Stage 4. Coefficient β_{1m} is estimated by regressing the outcome $Y_{ijm}{}^{*}$ on the resultant X -variables, correcting for measurement error.

Stage 5. Stages 2 through 4 are repeated a large number of times, in this example 500, and the mean β_1' of the β_{1m} estimated. The difference is then used as the next estimate of the bias and applied to the original β_0 to give the next estimate β_1

$$\beta_1 = \beta_0 - (\beta_1' - \beta_0)$$

Stage 6. The process of Step 5 is then repeated, starting with β_1, to give further estimates β_2, β_3, and so forth, until it converges to a value β_∞.

Convergence of the bootstrapping procedure was investigated graphically in each application to ascertain how many iterations were required. It was found that only one iteration was required to produce convergence, subsequent iterations being in the nature of oscillations. For this reason, the standard error is taken directly from the bootstrapping procedure.

The investigation

In all, six approaches to estimating the effect of measurement error on the results were used, and the results are now described. These were:

Program— the values estimated by the MLn program (program), and then *Resampling residuals* — the parametric bootstrap on observed X (parametric observed); conditional semiparametric (CSP), then *Resampling units* — a two-stage random resampling bootstrap (TSRR); resampling entire level-2 units (level-2 only), then *Bias-correction methods* — a bias-correction method based on that of Kuk (1995). In each case, resampling is with replacement.

DATA

The data came from a longitudinal study of school effectiveness in teaching reading in one Outer London Local Education Authority (LEA). Pupils were tested in 1988 and 1990 on the Southgate reading test (Hagley, 1987). Additional pupil- and school-level information was collected from schools. Complete information was available on 1,700 pupils (see Hutchison, 1998). Pupil-level variables to be included in the model were 1988 score, (Sscore88), sex, and free school meals. The aim of the research was to investigate the two school-level aggregated group-level variables (AGLEs) pupil turnover and mean school 1988 score. The outcome variable was 1990 score (Sscore90).

RESULTS

The program MLn used to estimate the error-corrected results was a beta version, and it was decided to use bootstrap procedures to verify or calibrate the results from it in two ways.

Bias Correction. In the first, a random error quantity was added to the x-variables used, data generated, and the error-correction MLn procedure was applied to estimate β^+. The rationale for this was that if the mean of the bootstrap results for $\hat{\beta}^+$ were not different from the original estimate, then it would seem reasonable that the MLn error-correction procedure was unbiased.

Estimation of Standard Error.

In the second verification procedure, the standard errors of the $\hat{\beta}^+$ bootstrap estimates were determined and compared with those from the program. In the end, it turned out that different bootstrap procedures gave different outcomes. An important feature of this chapter is the comparison of types of bootstrap techniques for multilevel techniques in general and especially for errors-in-variables.

Three representative values of level-1 reliability in (Sscore88), 1.00, 0.95, and 0.90, were investigated. It was assumed here that the other independent variables were measured without error at this stage. In order that the total effect of measurement error could be assessed, the first analyses in each set (reliability 1.00) did not allow for level-2 sampling error in mean Sscore88.

AGLEs, measurement error, and bootstrapping

This section deals with the question of aggregated group-level effects (AGLEs) and measurement error. Results are bootstrapped to compare with the program results. It was found elsewhere that a simple random resampling method was flawed in its application to this problem, and is excluded.

Estimates of the values of level-1 and level-2 error for a range of values of ρ_1, within-school reliability of 1988 Standardized Score, were fed into the equation in the program MLn (Rasbash & Woodhouse, 1995). Results for the program and bootstrap methods are shown in Tables 10.1 to 10.6. The last column in each table, labeled *Mean88*, shows the effect of the mean score. If this was statistically significant, then there was a statistically significant AGLE for the 1988 Standardized Score.

Level-1 fixed coefficients

Starting with level-1 fixed coefficients, it can be seen for the three values of the reliability coefficient considered, the results for coefficients for sex, free school meals, and score88 were substantially, although not identically,

TABLE 10.1

Program Multilevel Analysis: AGLEs 1990 Reading Score by 1988
Reliability

Fixed Coefficients and Standard Errors					
Rho	Sex	Free School Meals	Sscore88	Pupil Turnover	Mean88
1.00	-1.34	-2.37	0.79	-10.01	0.00
	(0.37)	(0.84)	(0.014)	(4.40)	(0.062)
0.95	-1.46	-2.18	0.83	-9.65	-0.03
	(0.37)	(0.85)	(0.016)	(4.48)	(0.067)
0.90	-1.60	-1.97	0.88	-9.27	-0.07
	(0.38)	(0.86)	(0.017)	(4.51)	(0.069)
Random Coefficients and Standard Errors					
	Level-2	**Level-1**			
1.00	2.38	56.59			
	(0.87)	(2.0)			
0.95	2.40	51.10			
	(0.87)	(2.0)			
0.90	2.44	45.00			
	(0.88)	(2.0)			

Note. Standard Errors in parentheses

equal for the various methods. All lie within 0.25 *SE* of the program result. For all bootstrap techniques, with increasing measurement error, the sex coefficient increased in magnitude, becoming increasingly negative with decreasing reliability. Conversely, the coefficient of free school meals become less negative, although with a blip in TSRR. The coefficient of Sscore88 increased with decreasing reliability.

The standard errors of level-1 coefficients were relatively unaffected by the introduction of measurement error with relatively small increases as error variances themselves increased, although there was an increase of the order of 20% in those of Sscore88. Sex coefficients were statistically significant at the .05 level. Free school meals coefficients were statistically significant for all analyses except TSRR and level-2, only because of the higher standard error obtained in these methods. The Sscore88 coefficients were statistically significant for all methods and all reliabilities.

Level-2 fixed coefficients

TABLE 10.2

Parametric Bootstrap Multilevel Analysis: AGLEs 1990 Reading Score by
1988 Reliability

Fixed Coefficients and Standard Errors						
Rho	**Sex**	**Free School Meals**	**Sscore88**	**Pupil Turnover**	**Mean88**	**Number Converged (of 500)**
1.00	-1.32	-2.37	0.79	-10.12	0.00	499
	(0.36)	(0.87)	(0.014)	(4.27)	(0.061)	
0.95	-1.46	-2.12	0.83	-9.11	-0.03	500
	(0.38)	(0.90)	(0.014)	(4.23)	(0.062)	
0.90	-1.59	-1.91	0.88	-8.71	-0.06	500
	(0.38)	(0.90)	(0.015)	(4.25)	(0.064)	
Random Coefficients and Standard Errors						
	Level-2	**Level-1**				
1.00	2.19	56.49				
	(0.86)	(1.9)				
0.95	2.22	51.05				
	(0.85)	(2.1)				
0.90	2.26	44.91				
	(0.85)	(2.1)				

Note. Standard Errors in parentheses

Among level-2 coefficients pupil turnover coefficients became less negative with decreasing reliability whereas the coefficients of mean Sscore88 decreased, becoming negative. Standard errors differed mainly between residuals and units resampling.

Although the pupil turnover coefficients became smaller, they remained statistically significant (greater than twice their standard errors) for the program, residuals resampling, and bias-correction methods. Because of the larger standard errors, none of the pupil turnover coefficients were statistically significant for the TSRR. The coefficient of mean Sscore88 went effectively from zero to negative: None of the values was statistically significant at the .05 level.

Random Effects

Results for level-2 variances were less consistent between methods in that the between-method differences were a larger proportion of the program value. All the bootstrap methods except the bias correction gave smaller

TABLE 10.3

Conditional Semi Parametric Multilevel Analysis: AGLEs 1990 Reading
Score by 1988 Reliability

Fixed Coefficients and Standard Errors						
Rho	**Sex**	**Free School Meals**	**Sscore88**	**Pupil Turnover**	**Mean88**	**Number Converged (of 500)**
1.00	-1.31	-2.32	0.79	-10.19	0.00	500
	(0.37)	(0.82)	(0.015)	(4.28)	(0.063)	
0.95	-1.44	-2.12	0.83	-9.73	-0.03	499
	(0.36)	(0.84)	(0.016)	(4.40)	(0.065)	
0.90	-1.58	-1.91	0.88	-9.21	-0.06	499
	(0.36)	(0.85)	(0.016)	(4.46)	(0.065)	
Random Coefficients and Standard Errors						
	Level-2	**Level-1**				
1.00	2.10	56.52				
	(0.82)	(2.2)				
0.95	2.11	51.06				
	(0.85)	(2.2)				
0.90	2.15	44.93				
	(0.88)	(2.2)				

Note. Standard Errors in parentheses

estimators for level-2 variance than did the program. Standard errors were comparable between methods, except that the TSRR results were clearly larger. Level-1 random coefficients all reduced substantially with decreasing reliability. Results were fairly close for all methods. Standard errors were larger for the two-stage resampling method, but all were statistically significant.

Comparing methods, it can be seen that TSRR, as employed in this section, by adjusting the level-1 and level-2 error corrections, has as many as 10% of the replications failing to converge satisfactorily. For this reason, it seems that other methods would be preferable.

Fixed coefficients are comparable between methods. Level-2 coefficients, especially pupil turnover, are apparently more variable between methods, but these coefficients are, in any case, less well determined because they are based on the relatively smaller number of level-2 units instead of those at level-1.

TABLE 10.4
Two-stage Non Parametric Bootstrap Multilevel Analysis: AGLEs 1990
Reading Score by 1988 Reliability

Fixed Coefficients and Standard Errors						
Rho	**Sex**	**Free School Meals**	**Sscore88**	**Pupil Turnover**	**Mean88**	**Number Converged (of 500)**
1.00	-1.29	-2.18	0.79	-11.05	0.00	493
	(0.50)	(1.56)	(0.024)	(5.76)	(0.098)	
0.95	-1.42	-2.00	0.83	-10.43	-0.03	467
	(0.50)	(1.57)	(0.025)	(5.74)	(0.098)	
0.90	-1.56	-2.14	0.88	-9.44	-0.06	451
	(0.51)	(1.58)	(0.027)	(5.61)	(0.107)	
Random Coefficients and Standard Errors						
	Level-2	**Level-1**				
1.00	1.61	55.78				
	(0.99)	(3.8)				
0.95	1.85	50.38				
	(1.03)	(3.8)				
0.90	2.04	44.73				
	(1.15)	(3.8)				

Note. Standard Errors in parentheses

COMPARING METHODS

Level-1 variances are comparable between methods. In comparison to their size, level-2 variances are more variable between methods, with the TSRR especially different. The standard errors for all coefficients are comparable for all methods except TSRR, which are of the order of 50% larger in some instances, except for l-2. Level-2 only is intermediate. It seems that the sampling of units rather than residuals gives greater variation in comparison with other methods.

Standard errors are comparable for level-1 coefficients for the program and resampling residuals methods (parametric and CSP), whereas the units resampling (two-stage) has substantially larger standard errors. Level-2 only is intermediate, with standard errors being substantially larger for free school meals and Sscore88. Level-2 only results are comparable with the rest for sex, pupil turnover, mean88, and level-2 random coefficients.

Only one correction step is required for the bias-correction method for

TABLE 10.5

Level-2 Units Resampling Multilevel Analysis: AGLEs 1990 Reading
Score by 1988 Reliability

Fixed Coefficients and Standard Errors						
Rho	**Sex**	**Free School Meals**	**Sscore88**	**Pupil Turnover**	**Mean88**	**Number Converged (of 500)**
1.00	-1.34	-2.36	0.79	-9.66	0.01	499
	(0.36)	(1.24)	(0.018)	(4.65)	(0.069)	
0.95	-1.47	-2.13	0.83	-9.46	-0.02	495
	(0.39)	(1.30)	(0.019)	(4.59)	(0.063)	
0.90	-1.61	-1.91	0.88	-9.10	-0.06	491
	(0.40)	(1.32)	(0.020)	(4.63)	(0.064)	
Random Coefficients and Standard Errors						
	Level-2	**Level-1**				
1.00	2.24	55.88				
	(0.92)	(2.9)				
0.95	2.20	50.65				
	(0.97)	(3.1)				
0.90	2.23	44.54				
	(0.97)	(3.1)				

Note. Standard Errors in parentheses

all the values of measurement error considered here.

CONCLUSIONS

Two main topics have been considered here: First, the effect of
measurement error in scores as independent variables, and secondly, the
possibilities of bootstrapping errors-in-measurement multilevel models.
The original aim of bootstrapping the results was as a verification or
calibration of the MLn results for bias and variation. However, it was
found that different bootstrap results in fact give different outcomes, and
an important feature of this chapter is thus a comparison of a variety of
bootstrapping approaches.

Methodological findings

The standard deviations and biases of the coefficients were investigated
by using bootstrap techniques. Perhaps one of the most important findings
is that to refer to the bootstrap is to risk prompting a misleading inference,

TABLE 10.6

Bias-Correction: 1 Iteration Multilevel Analysis: AGLEs 1990 Reading
Score by 1988 Reliability

Fixed Coefficients and Standard Errors						
Rho	**Sex**	**Free School Meals**	**Sscore88**	**Pupil Turnover**	**Mean88**	**Number Converged (of 500)**
1.00	-1.35	-2.38	0.79	-9.89	0.00	500
	(0.36)	(0.87)	(0.014)	(4.27)	(0.061)	
0.95	-1.46	-2.24	0.83	-10.18	-0.03	500
	(0.38)	(0.90)	(0.014)	(4.23)	(0.062)	
0.90	-1.60	-2.02	0.88	-9.83	-0.07	499
	(0.38)	(0.90)	(0.015)	(4.25)	(0.064)	
Random Coefficients and Standard Errors						
	Level-2	**Level-1**				
1.00	2.58	56.68				
	(0.86)	(1.9)				
0.95	2.58	51.16				
	(0.85)	(2.1)				
0.90	2.62	45.09				
	(0.85)	(2.1)				

Note. Standard Errors in parentheses

as different bootstrap methods gave different results. Methods based on units bootstrapping and residuals bootstrapping are considered in turn.

In the units bootstrap, the random aspects of the model, namely the measurement error and the residuals, are incorporated into the observed data, and in resampling the units, one is at the same time creating a model distribution of the random aspects. Because the random elements come packaged with the fixed elements in a units bootstrap, this means that any relationship between the fixed and random elements exhibited in the original sample will also be shown in the bootstrap.

Two units resampling methods were tried, two stage (TSRR), and level-2 only. (A simple random resampling procedure was tried elsewhere, and found to give biased results.) The possibility of using a units resampling method to mimic the within-level-2 sampling that is part of level-2 error variance was considered, but was found not to be feasible.

Two residuals resampling methods were also investigated. These were a parametric (residuals) bootstrap, and a procedure using the observed residuals. In general, all bootstrapping procedures tend to give comparable results for the fixed coefficients. However, the position is more complex for the random coefficients.

The results for all the residuals resampling methods were found to be quite close to those of the program. In general, this is rather encouraging, because the CSP did not rely on assumptions of normality in its residuals in its resampling. Among the units resampling methods, a level-2 only resampling was quite successful. Other units approaches were biased, and required procedures to correct for this. There is a tendency for units resampling to give higher values of standard errors than residuals resampling methods.

Rather than ad hoc attempts at allowing for biases in the bootstrapping procedures, a more general process may be preferable, and one based on that associated with Kuk (1995) was used. It was found that at most, one iteration was required for convergence.

Because the cases where the estimates from the various bootstrap procedures differ from the model and each other tend to be explicable in terms of specific biasing features of the process, this is rather encouraging about the values of the coefficients and their standard errors provided by the program MLn. Among the bootstrapping procedures investigated, the parametric, the level-2 only and the bias-correction procedures were particularly successful and simple to implement.

Substantive findings

Apparent group-level effects of a variable can disappear when allowance is made for measurement error in that variable. However, allowing for measurement error in a variable can have wide-ranging effects, also substantially reducing apparent group-level effects in other variables.

Footnote

An alternative strategy, in which schools are selected with probability proportional to size, was considered, but not chosen because it would have given a set of pseudo-schools all of equal size, which seemed to be getting away from the structure of the data.

REFERENCES

Carpenter, J., Goldstein, H., & Rabash, J. (1999). A non-parametric bootstrap for multilevel models. *Multilevel Modelling Newsletter, 11* (1), 2–5.

Draper, N. R., & Smith, H. (1981). *Applied regression analysis*. New York: Wiley.

Fuller, W. A. (1987). *Measurement error models*. New York: Wiley.

Goldstein, H. (1979). Some models for analysing longitudinal data on educational attainment. *Journal of the Royal Statistical Society, 142* (3), 407–442.

Goldstein, H. (1995). *Multilevel statistical models* (2nd ed.). London: Arnold.

Hagley, F. (1987). *Suffolk Reading Scale: Teacher's guide*. Windsor: NFER-NELSON.

Haukka, J. M. (1995). Correction for covariate measurement error in generalised linear models — A bootstrap approach. *Biometrics, 51*, 1127–32.

Hutchison, D. (1998). The effect of group-level influences on pupils' progress in reading. Unpublished doctoral dissertation, University of London.

Kuk, A. (1995). Asymptotically unbiased estimation in generalized linear models with random effects. *Journal of the Royal Statistical Society*, (2, Series B), 395–407.

Meijer, E., Van Der Leeden, R., & Busing, F. M. T. A. (1995). Implementing the bootstrap for multilevel models. *Multilevel Modelling Newsletter, 7* (2), 7–11.

Mooney, C. Z., & Duval, R. D. (1993). Bootstrapping: A non-parametric approach to statistical inference. *Quantitative Applications in the Social Sciences* (No. 95). Newbury Park, CA: Sage.

Rasbash, J., & Woodhouse, G. (1995). *MLn Command Reference:* (Version 1.0). Multilevel Models Project, Institute of Education, University of London.

Wacliwiw, M. A. & Liang, K. Y. (1994). Empirical Bayes estimation and inference for the random effects model with binary response. *Statistics in Medicine, 13*, 541–51.

Woodhouse, G., Yang, M., Goldstein, H., Rasbash, J. & Pan, H. (1996). Adjusting for measurement error in multilevel analysis. *Journal of the Royal Statisitcal Society, 159* (2, Series A), 201–212.

11

An Iterative Method For the Detection of Outliers In Longitudinal Growth Data Using Multilevel Models

Russell Ecob
Geoff Der
MRC Social and Public Health Sciences Research Unit,
4 Lilybank Gardens, Glasgow G12 8RZ

The treatment of outliers has a long history. Many of the most influential statisticians have made contributions to the detection and treatment of outliers in data, their work spanning over a century, often devoting substantial portions of their career to this study (details can be found in excellent reviews of this field; Barnett & Lewis, 1994; Hawkins, 1980). More recently, multilevel modelling methods have been used to diagnose outliers with theoretical work on diagnostics in multilevel models with particular application to the detection of outliers (Hilden-Milton, 1995; Hodges, 1998), and a number of practical procedures for dealing with outliers, applied to cross-sectional educational data (Langford & Lewis, 1998). However, although there have been many recent examples of the application of multilevel and related growth models to longitudinal data, both on physical attributes and psychological processes, with both single and multivariate dependent variables (Goldstein, 1989, 1995; Hoeksma & van der Boek, 1993; Longford, 1993; Rogosa & Saner, 1995; Sayer & Willett, 1998; Ware, 1985), and although longitudinal missing data has received some attention (Diggle, 1994; Diggle & Kenward, 1998; Goldstein, 1995), to our knowledge no application of multilevel modeling methods has been made to the detection of outliers in longitudinal data.

Standard statistics texts on outliers provide us with the following definitions of outlier. Hawkins (1980, p. 1) gave an "intuitive" definition as "...an observation that deviates so much from other observations as to arouse suspicion that it was generated by a different mechanism." Barnett and Lewis (1994, p. 7) provided the more formal definition as "an observation (or a subset of observations) which appears to be inconsistent with the remainder of that set of data." Coining the concept of discordancy, they stated "...an observation is discordant when it is statistically unreasonable (as indicated through a statistical testing procedure) on the basis of a prescribed probability model for the data." The definitions of Hawkins and of Barnett and Lewis are related in the equivalence of the prescribed probability model (Barnett & Lewis) and the generating mechanism (Hawkins). The necessarily subjective nature of the outlier detection process is indicated through the use of the words "to arouse suspicion" and "appears to be inconsistent." The dependence of the outlier on the supposed generating mechanism or probability model for the data is illustrated by the following. Consider the maximum of a positively skewed distribution. If the generating mechanism is the log-normal distribution, such an observation may well be consistent with the generating mechanism. However, if the supposed generating mechanism operates in a way such as to give rise to a normal distribution, it is likely to be considered as discordant and an outlier. The issue to which this chapter is addressed is the effect of the outlier(s) on the elicitation from the data of the generating mechanism and the minimization of this effect.

In multilevel data, we have the complication that outliers may occur at more than one level.[1] For example, in data on progress of pupils in schools, particular schools may achieve substantially above or below expectation given intake characteristics. Alternatively, particular pupils may be recorded as making particularly good or bad progress. Errors of measurement or testing of a particular pupil will distort the estimate of the

---

[1]Let us consider repeated measures data on individuals. It is standard practice to consider the repeated measures, at level 1, as nested within the individuals, at level 2. Outliers at a higher level in the data may result from a particular "rogue" interviewer or observer who misread instructions or to an overall misreporting by a measuring instrument and may be best detected through the inclusion of interviewer as a random factor at the level above the individual in a three–level model. At the lowest level, outliers may result from a particular mispositioning or misreading of the measuring instrument or from a transcription error in the recording procedure. The general overestimation of a quantity, say height, by a particular interviewer may be counteracted by a particular observation that is underrecorded, possibly by a misreading of a scale.

effect of the particular school of which he or she is a part. Conversely, a crude model for pupil progress that contains no information on the school could lead to progress of particular pupils being inappropriately considered as outliers when the school is over or underperforming. Similarly, for growth data, an individual with greater than expected estimated growth over a particular follow-up period may be either the result of a real difference in growth, possibly due to either delayed maturation or greater achieved adult height, or as a result of an incorrectly recorded observation, particularly near one of the extremes of the age range considered.

Any procedure for the detection and subsequent deletion of outliers in multilevel data therefore needs to take the multiplicity of levels at which they might occur into account. Given the reasonable restriction to a procedure that operates at one or more levels sequentially, what temporal priority should be given to the level at which outliers are screened? Langford and Lewis (1998), on cross–sectional data on schools, argued for priority at the "higher" school level for the following reasons;

1. Researchers are often most interested in the higher level of aggregation.
2. "If discrepancies can be found in higher level structures these are more likely to be indicative of serious problems than a few outlying points in lower levels are" (p. 124).

For the longitudinal data examined here, we argue for priority at the level of wave or occasion of measurement within the individual. This is for the following reasons:

1. An outlier at a specific wave distorts estimates of the individual level parameters.
2. The multilevel models allow for differences between individuals over all waves. Given a reasonable model for the data, this generally precludes the incorrect attribution of differences between individuals to specific waves. Thus, outliers detected at specific waves are normally correctly attributable to the wave in question.
3. Individuals are generally measured at different waves by different observers or interviewers — thus any errors of measurement, including those related to a particular interviewer (Ecob & Jamieson, 1992) are likely to be occasion specific.
4. Longitudinal data typically have few observations within each individual, thus increasing the potential for outlying observations at particular waves to have effects on estimates of residuals corresponding to particular individual level parameters (e.g., mean, slope). In contrast, cross-sectional data typically have a larger number of individuals in each higher level unit (e.g., school). The effect of an incorrect observation on the estimate for the higher level unit is therefore smaller.

In choosing to opt for the lowest level as priority, we concur with Raab and

Parpia in the discussion in Langford and Lewis (1998).

The problem with detecting outliers or discordant observations through fitting a statistical model to the whole data is that these discordant observations, when not correctly modeled, influence the choice of the model used to detect them. In this chapter, we aim to get around this problem through an iterative method of detecting outliers in longitudinal data comprising repeated measures on individuals over time using multilevel models. This is based on the supposition that the outlier — often an incorrect observation — will distort a statistical model fitted to the data. This will, in turn, have effects on the evaluation through the residuals from the fitting of the model, of the correctness, or otherwise of the remaining observations. By iterating, the identification of the error-prone observations is improved (see Anscombe, 1960).

How does a multilevel approach to screening for outliers compare with a naïve screening method? A naïve method would screen separately for outliers at each wave. This would take no account of legitimate differences between individuals, an individual toward the top end of the height distribution giving rise to outliers at each wave, all detected by the naïve method. Note also that the naïve method also would not detect or eliminate an observation that is not extreme on a given wave, even when it is the result of an (say downward) error for a particular wave of an individual with generally high values. However, by allowing for individual differences through incorporating a random error at the individual level in a multilevel model, extreme individuals no longer give rise to extreme observations at the level of the wave within the individual. Thus, the proposed multilevel method only eliminates the observation when extreme at a particular wave or waves — if extreme at all waves this contributes to the error at the individual level and may be separately modeled by a dummy variable corresponding to the individual or deleted at a later stage (the naïve method makes no such distinction between the waves and the individual, and eliminates these observations in all cases). The identification, and possible deletion from the data, of extreme individuals thus follows a process of identification of extreme observations at particular waves relative to the observations for an individual overall. The major difference between the two methods is one of philosophy — screening on the basis of the multilevel model allows estimation and screening to be integrated in the same procedure. Better detectability of outliers (both of sensitivity and selectivity) should result.

Our method for the outlier detection is assessed on two datasets. The first dataset comprises heights of a cohort of 15-year-olds (males and females) subsequently followed up at ages 16, 18, and 23 from a study examining the social patterning of health over time [West of Scotland twenty-07 study (Macintyre et al., 1989)]. On this data we:

1. examine the sensitivity of the parameters estimated from the model to the particular value of a tuning parameter, the cut-off above which observations are temporally eliminated at each iteration.

2. assess (for females only) the ability of the procedure to recover the original model when the heights of a random subset of the data on a particular wave are perturbed by a fixed amount.

The second dataset (Kenward, 1987) is of the weight of calves on 11 occasions for two groups comprising different treatments, the experimental treatment being assessed for its influence on weight gain in comparison to a control treatment. This data illustrates the applicability of the method to data collected on a larger number of occasions. For this data, we compare the relative effect of the iterative residual deletion procedure and the misspecification of the multilevel model through the unnecessary simplification of either or both the random and fixed parts.

For both data sets, these models allow for nonlinear relationships of the measure to the wave or occasion at which it was measured. Polynomial models used here have been found to be useful ways of representing this data (Goldstein, 1989). The models include separate terms for powers (e.g., quadratic, cubic) of the difference of the age or occasion from a specified age or occasion. They have advantages over nonlinear (for example, exponential) models in that these models are linear in the parameters and so can be estimated using the standard multilevel modeling methods.

Finally, we note that the outlier detection and deletion methods were used on the height data as part of the cleaning process, whereas we make the assumption that the calf data has already been cleaned. We would therefore expect a larger proportion of the height data to be identified as outliers.

THE PROPOSED METHOD

The proposed procedure for detection and subsequent deletion of outliers involves an iterative application of empirical Bayesian estimates using a multilevel model (Goldstein, 1995). These models are fitted with waves nested within the individual and allow in principle for a range of individually specific random parameters (for intercept, slope, etc). These random parameters can be allowed to vary by further factors (e.g., social groups). An incorrect observation will distort a model fitted to the whole data thereby biasing the evaluation of the correctness or otherwise of the remaining observations. By iterating, we can improve on the identification of the error-prone observations.

The iterative process proceeds as follows. First a model is fitted to the whole data. Then, observations that are judged to be outliers, based on residuals above a specified criterion or cutoff, are temporally deleted from the data set and the model refitted. The complete data is then examined in the context of the updated model and large residuals again temporally deleted, and the model again refitted. This procedure repeats until convergence. This is achieved when the set of data points judged to be outliers does not change between iterations. By definition, the parameter estimates will have then also converged. Only residuals at the lowest level are evaluated in this procedure. However, this method can be easily extended to examine residuals at higher levels (individuals and above).

Models

The multilevel models used in this chapter are of the form

$$y_{ij} = \beta_{0j} + \beta_1 x_{ij} + \ldots + \beta_k x_{kj} + e_{ij}, \tag{11.1}$$

where i denotes the wave or occasion within the individual and j the individual. The explanatory variables $x_{ij} \ldots x_{kj}$ include variables specific to the individual such as polynomial terms in age or occasion, sex, and treatments where relevant, as well as possibly wave or occasion specific variables (for example, dummy variable to denote self-report). For the intercept to have meaningful interpretation and better computation, all terms involving age or occasion need to be centered by subtracting a particular value before raising it to the required power. For the calf data, for example, measured in occasions, 5 is subtracted from all terms so that the intercept refers to the weight at occasion 5. This also means that the estimated effect of the difference between treatment groups is at this occasion. The height data are centered at the average age at first wave.

In this model, only the individual level intercept term (β_{0j}) varies randomly, as follows:

$$\beta_{0j} = \beta_0 + u_{0j}. \tag{11.2}$$

For this "simple random structure" model (used in one of the calf data models), all random errors are assumed to be normally distributed and to have zero expectation, and to (co)vary as follows;

$$
\begin{aligned}
Var(e_{ij}) &= \sigma_e^2 \quad \text{(constant across waves)} \\
Var(u_{0j}) &= \sigma_1^2 \\
Cov(e_{ij}, e_{ik}) &= 0 \quad \text{for individuals} \quad j, k, j \neq k \\
Cov(e_{i1j}, e_{i2j}) &= 0 \quad \text{for waves} \quad i_1, i_2, i_1 \neq i_2 \\
Cov(u_j, e_{ij}) &= 0
\end{aligned}
$$

Further models tested allow the first of these assumptions to be relaxed, allowing variation, or heterogeneity, of errors by wave; thus, $Var(e_{ij}) = \sigma_{ei}^2$. This occurs, for example, when self-reported height, on wave 2 only, is biased in relation to measured height, this bias varying randomly according to the individual. Heterogeneity of variance is also found for the calf data, the variance varying according to the square of the occasion. The fourth assumption would be reasonable when waves or occasions are reasonably far apart in time. If not, the covariance can be modeled as having one of a variety of autocorrelation structures (Goldstein, 1995, pp. 91–92).

Multilevel models can be further generalized to allow any of the coefficients (e.g., β_1) to vary according to the individual (becoming β_{1j}) with

$$\beta_{1j} = \beta_1 + u_{1j}. \tag{11.3}$$

When coefficients β_{k1j}, β_{k2j} vary in this way, the corresponding errors u_{k1j}, u_{k2j}, at the individual level will then generally covary as follows:

$$Cov(u_{k1j}, u_{ik2j}) = \sigma_{k1k2}^2. \tag{11.4}$$

It is empirically more likely that the polynomial coefficients of lower degree (i.e., linear) will be found to vary than those of higher degree (i.e., quadratic). This variation can in turn, of course, be related to further characteristics of individuals (for example, social class, if measured). Such variation occurs in the random coefficient models used in the calf data.

Multilevel models are particularly appropriate for longitudinal data (as with the height data) with attrition over waves and with the times of measurement varying within the specific wave, as they are efficient, making use of all the available data. When data is missing, the assumption that it is "missing at random" is necessary for the resulting estimates to be unbiased. When it is considered that the missingness of the data may be related to the (hypothetical) dependent variable value (informative dropout), then parameter estimates from such procedures will be biased and other methods should be used (Diggle & Kenward, 1994; Diggle, 1998).

Debates regarding the usefulness of polynomial models and alternatives can be found in Royston and Altman (1994). A possible improvement to the standard polynomial modeling procedures is the grafted polynomial multilevel model, which fits separate polynomial models to different sections (sets of waves) of the data, constraining the manner and rates of change of the different polynomials at the graft points to be equal. An application of such a model to growth data was given by Pan, Goldstein, and Di (1992). All analyses were carried out using MlwiN (Rasbash et al., 2000).

RESULTS

Height data

This data set came from the youngest cohort of the West of Scotland twenty-07 study—health in the community (Macintyre et al., 1989), and was comprised of measurements of height at four waves at ages 15, 16, 18, and 23. Respondents were randomly allocated to interviewers (trained nurses) in a constrained random fashion allowing each to interview in particular areas (see Ecob & Jamieson, 1992 for details). Measurement was obtained from a portable stadiometer in the person's own home for waves 1, 3, and 4. For wave 2, a postal survey replaced the interview. A physical measure of height was therefore not possible and a self-report was used as a proxy. Interviews on a given wave were spread over a period that varied according to the wave, being larger in the final wave (a spread of around 6 months) than in the earlier waves (spread of 3–4 months). The date of interview was recorded and used to calculate age at interview, measured in months. Some attrition was present in this data, waves 1 through 4 having respectively 950, 794, 897, and 670 observations. Note in particular the additional nonresponse at wave 2 due to the postal survey.

For this data set, males and females were modeled separately. Age at interview (centered at 15 years, 8 months) and date of birth were included in all models. Any bias in height caused by the self-report in relation to the measured height (there is some evidence from other sources that people overreport their own height) was allowed for by including a dummy variable to represent the difference between self-report and hypothetical measured report. This difference was allowed to vary randomly by individual, such random variation being found for females but not for males (with variance 1.97 estimated with standard error of 0.57). Table 11.1 shows the correlations of the height measures across waves, separately for males and females. The correlations for females were similar across waves and were higher than for males in the earlier waves. This reflects the greater degree to which the females attained their adult height at these ages.

Table 11.2 shows, separately for males and females, estimates for the fixed part (only) for the original model and the final model, after deletion of outliers. The coefficient of date of birth (not shown) failed to reach statistical significance, indicating that the variation with age between waves corresponds to that within. In this table, the effects of varying the cut-off criterion are shown. Starting with the females, the effect of the deletion of outliers is to increase the linear coefficient of age and, to a larger extent in absolute terms, the quadratic coefficient for age. The percent increase in the estimate of rate of change of height with age at 15 years for the three cut-off values of 2, 3, and 4 standard deviations was 36%, 15%, and

TABLE 11.1
Correlations of Heights Across Waves, by Gender

	Wave 1	*Wave 2*	*Wave 3*
Males			
Wave 2	0.81		
Wave 3	0.76	0.83	
Wave 4	0.70	0.81	0.90
Females			
Wave 2	0.89		
Wave 3	0.87	0.88	
Wave 4	0.91	0.91	0.90

Note. Only data that was common across all waves is included in these correlations. This restriction is not necessary for the multilevel modeling, which uses all available data.

15%, respectively. The estimate of the dummy variable for wave 2 was reduced. For males, in contrast, the effect of deletion of outliers was to decrease both linear and quadratic coefficients of age; the percent decrease for the three cut-off values 2, 3, and 4 standard deviations was 19%, 8%, and 5%, respectively. The estimate of the dummy variable for wave 2 was again reduced but by a lesser amount than for females. Note the much larger coefficient of age for males, consistent with their later maturity than females, and also the larger estimate of the self-report dummy variable for males.

The fitted data (excluding the dummy variable for wave 2) for females is shown in Fig. 11.1. The plots are for ages in the range of 180 to 260 months, which comprise 80% of the data.

The basic model is shown by the continuous line and that corresponding to the 4, 3, and 2 standard deviation cutoffs by dashed lines of correspondingly shorter lengths.

The fitted curves after the deletion of outliers are generally above the original for waves 2 and 3, but below for wave 1. The curves for the cutoffs of 3 and 4 standard deviations are similar and less different from the original curve than is that for the cutoff of 2 standard deviations.

For males, corresponding plots are shown in Fig. 11.2. Here, the fitted curves after the deletion of outliers are generally above the original for wave 1 but below for wave 3 and show differences in the direction opposite to the females. Again, the curves corresponding to the cutoffs of 3 and 4 standard deviations are more similar to each other (and to the original)

TABLE 11.2

Height Data Showing the Effect of Variation in Cut-off Criterion for
Nonperturbed Data

Original Model				
Criterion (SD) by Sex	Age (Linear)	Age (Quadratic) $*10^{-4}$	Wave 2 Dummy Variable	No. of Iterations
Males				
2	0.170 (0.008)	-12.00 (0.79)	2.50 (0.20)	3
3	" "	" "	" "	1
4	" "	" "	" "	4
Females				
2	0.019 (0.006)	-1.02 (0.52)	1.03 (0.13)	3
3	" "	" "	" "	2
4	" "	" "	" "	2
Final Model				
Males				
2	0.138 (0.006)	-9.80 (0.54)	2.29 (0.14)	3
3	0.155 (0.007)	-11.00 (0.65)	2.37 (0.16)	1
4	0.162 (0.007)	-11.48 (0.71)	2.49 (0.18)	4
Females				
2	0.026 (0.004)	-1.81 (0.36)	0.65 (0.09)	3
3	0.022 (0.004)	-1.34 (0.41)	0.80 (0.10)	2
4	0.022 (0.005)	-1.35 (0.44)	0.87 (0.11)	2

Note. Standard errors are in parentheses.

than is that corresponding to 2 standard deviations. Note that the rate of
change of height with age at age 15 from the original model is estimated at
2.0 centimeters per year for males and 0.23 centimeters per year for females.

At higher ages (not shown), although the mean fitted values in the fourth
wave are above the third wave for both sexes, the curves show a down turn
within the fourth wave. This probably results from the global nature of the
fit of the polynomial modeling methods and is exacerbated by the much
larger gap between the third and fourth waves (average age difference, 5
years 7 months) than between the previous waves (1 year 3 months and 1
year 5 months on average between waves 1 and 2 and between waves 2 and
3 respectively). Grafted polynomial fitting (mentioned earlier) would be a
possible solution to this problem.

The apparent reduction in fitted height at the higher ages, particularly

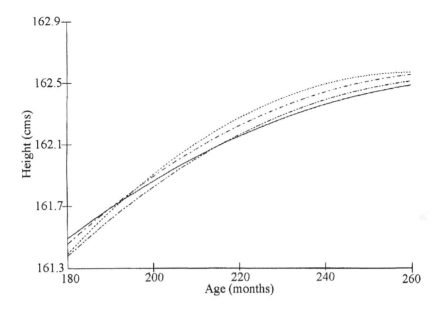

FIG. 11.1. Height data for females: Comparison of plots for original model and model after residual deletion using three alternative cutoffs — nonperturbed data.

for males, probably results from the global nature of the fit of the polynomial modeling methods, the pattern of change within wave 4 being largely determined by the fitting of the polynomial at the earlier waves comprising the majority of the data, and of the variation in the dependent variable. Alternative methods of modeling this data were mentioned earlier. This particular problem in fitting is exacerbated by there being a much larger age gap between the 3rd and 4th waves (average age difference between waves 1 and 2 and between waves 2 and 3 is 1 year 3 months and 1 year 5 months, respectively) than in previous waves (1 year 3 months and 1 year 5 months average age difference between waves 1 and 2 and between waves 2 and 3 respectively).[2]

As expected, the random variance at level 1 is reduced after iteration but the variance at level 2 is little affected (tables available from authors). A greater percentage of the data is eliminated toward the bottom end of

---

[2] Ranges of ages (in years and months) are, in waves 1 to 4 respectively, 15.0–16.10, 16.7–17.10, 17.11–19.4, 23.4–25.7.

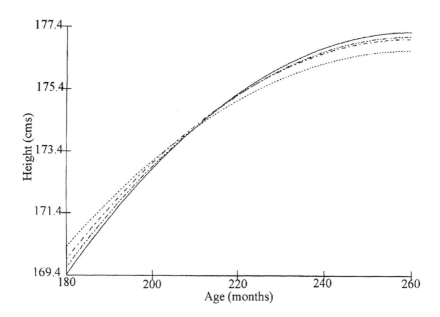

FIG. 11.2. Height data for males: Comparison of plots for original model and model after residual deletion using three alternative cutoffs — nonperturbed data.

the distribution of level 1 residuals for males but this is similar at both extremes for females. The percent eliminated, except for the cutoff of 2 standard deviations, is markedly larger than the percent estimated on the assumption of a normal distribution, consistent with the marked negative skewness of residuals at level 1. If this is not taken into account, standard errors of estimates of model parameters will be overestimated. This is indicated in Table 11.2, the final model giving lower standard errors even though the size of the sample on which the model was estimated, is reduced.

Table 11.3 shows the skewness and kurtosis of the original data by wave for each sex. The values are relatively small and variable between waves. However (Table 11.4), the skewness and kurtosis of the residuals from the original model tends to be much larger. Note in particular the much larger values of kurtosis for males at wave 2 and for females at wave 3, indicating some obvious data inaccuracies not so easily detectable through the plots of original data by wave and sex.

This data (for females only) is now modified by perturbing a percentage

TABLE 11.3

Skewness and Kurtosis Measures by Wave — Height Data

Wave	1	2	3	4
Males				
Skewness	-0.03	-0.26	0.16	0.16
Kurtosis	0.59	0.79	-0.30	-0.08
Females				
Skewness	0.12	0.13	-0.04	0.08
Kurtosis	0.17	0.29	0.29	0.39

Note. The skewness and kurtosis measures used are the measures γ_1, γ_2 being ratios of cummulants raised to appropriate powers for invariance to the scale measurement of the data (Kendall & Stuart, 1969).

of the observations at a particular wave, increasing these by a fixed amount (either 1 or 2 within wave standard deviations).[3] The parameters of the resulting final models are then compared to the original models both on the complete perturbed data and on the data before perturbation in relation to the "no error" model shown at the bottom of the table. A measure of the success of the method is the ability to reproduce estimates for final models that are similar to those on the unperturbed data. This would relate to the sensitivity and selectivity with which the perturbed observations are eliminated (i.e., what proportion of perturbed observations are correctly eliminated and what proportion of nonperturbed observations are incorrectly eliminated). In practice, the success would depend on the percentage of observations perturbed, the amount of perturbation, and the proportion of observations deleted from the data (via the tuning parameter) at each iteration of the method.

Tables 11.5 and 11.6 show, for females only, the effects on the parameter estimates in the fixed part of the model of a range of possible perturbations to the data. In all these analyses, the cutoff is fixed at 3 standard deviations. Perturbations vary according to the wave for which the data is perturbed (waves 1 to 4), the percentage perturbed on the particular wave (1%, 5%), and the amount of perturbation (1, 2 standard deviations respectively for Tables 11.5 and 11.6).

Let us consider first data for which 5% of observations on height are perturbed with 1 standard deviation added (bottom section, Table 11.5). The effect of the perturbation varies as expected according to the wave

---

[3]This is a special case of a contaminating distribution considered by Dixon (1950).

TABLE 11.4

Skewness, Kurtosis and Standard Deviations From Original Model by
Wave (Height Data)

Wave	1	2	3	4	Overall
Males					
Skewness	-1.31	-1.87	0.25	0.75	-0.79
Kurtosis	2.87	15.11	1.76	4.98	5.32
SD		3.00	2.31	2.21	2.52
Females					
Skewness	0.10	1.00	-3.56	-0.23	-0.72
Kurtosis	5.14	3.47	34.35	5.39	13.56
SD		1.06	1.97	1.83	1.58

— at wave 1, perturbation substantially reduces the linear and quadratic terms in age. In contrast, at wave 3, both linear and quadratic terms are increased in absolute value. The dummy variable coefficient is also affected, being increased as expected when perturbation is on wave 2 and reduced when on waves 1 and 3. Final estimates after deletion of large residuals gives, in absolute terms, generally larger linear and quadratic age terms with values usually nearer (although not when perturbation is at wave 3) the final estimates for the unperturbed (no error) data. When 1% of the data is perturbed in the same way, biases in estimates are substantially lower but in the same direction as for the 5% perturbation, and these are reduced for perturbations at all waves (relative to the no error model) in the final models. In particular, the linear coefficient of age and the coefficient of the wave 2 dummy variable are very similar in all final models.

Data for perturbation of 2 standard deviations according to the same scheme are shown in Table 11.6. The greater size of perturbation allows the method to more effectively discriminate and thus eliminate the perturbed data. Although the original model for perturbed data is further from that for the no error data as expected, the final models are generally more similar to each other and to the nonperturbed data model — for example, biases in all parameters shown when 1% of observations are perturbed have been almost completely eliminated. In all cases, the number of iterations is three or lower. Plots of the effect of perturbation of 5% of the data at 2 standard deviations at wave 3 are shown in Fig. 11.3.

Here, the long dashed line — generally the lowest in the plots — represents the original data before deletion of high residuals, and the continuous line represents the original data after residual deletion. The

TABLE 11.5

Height Data for Females; 1%, 5% Observations Perturbed Upward on One
Wave by 1.00 *SDs*

1% Observations Perturbed						
Model	Original			Final		
Wave[a]	Age (Linear)	Age (Quadratic) $*10^{-4}$	Wave 2 Dummy Variable	Age (Linear)	Age (Quadratic) $*10^{-4}$	Wave 2 Dummy Variable
1	0.016	-0.80	1.00	0.020	-1.26	0.79
	(0.006)	(0.53)	(0.13)	(0.004)	(0.41)	(0.10)
2	0.018	-1.00	1.11	0.022	-1.35	0.83
	(0.006)	(0.53)	(0.13)	(0.004)	(0.41)	(0.10)
3	0.020	-1.12	1.03	0.022	-1.38	0.79
	(0.006)	(0.52)	(0.13)	(0.004)	(0.41)	(0.10)
4	0.018	-0.89	1.04	0.022	-1.27	0.80
	(0.006)	(0.53)	(0.13)	(0.004)	(0.41)	(0.10)
No error	0.019	-1.02	1.03	0.022	-1.34	0.80
	(0.006)	(0.52)	(0.13)	(0.004)	(0.41)	(0.10)
5% Observations Perturbed						
Model	Original			Final		
Wave[a]	Age (Linear)	Age (Quadratic) $*10^{-4}$	Wave 2 Dummy Variable	Age (Linear)	Age (Quadratic) $*10^{-4}$	Wave 2 Dummy Variable
1	0.001	-0.07	0.85	0.011	-0.50	0.70
	(0.006)	(0.58)	(0.14)	(0.005)	(0.44)	(0.11)
2	0.018	-1.00	1.36	0.021	-1.19	1.00
	(0.006)	(0.56)	(0.13)	(0.005)	(0.43)	(0.11)
3	0.030	-2.14	0.88	0.031	-2.26	0.68
	(0.006)	(0.54)	(0.13)	(0.005)	(0.43)	(0.11)
4	0.017	-0.58	1.04	0.021	-0.99	0.85
	(0.006)	(0.54)	(0.13)	(0.005)	(0.43)	(0.10)
No error	0.019	-1.02	1.03	0.022	-1.34	0.80
	(0.006)	(0.52)	(0.13)	(0.005)	(0.41)	(0.10)

[a] wave on which perturbed observations are added

Note. Standard errors in parentheses.

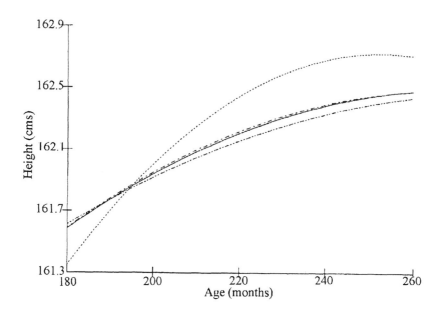

FIG. 11.3. Height data for females: Comparison of plots for original model and for model after residual deletion for a) 5% of data perturbed at wave 3 upward by 2 standard deviations and b) for nonperturbed data.

corresponding shorter dashed lines represent the data after perturbation, before (shortest dashes) and after (intermediate length dashes) residual deletion. The effect of deletion of residuals for the perturbed data is thus in this case to bring the relationship in line with the non-perturbed data.

Calf data

This data set was originally analysed by Kenward (1987). It comprised weights of 60 young calves each measured on 11 occasions[4] throughout the grazing season at 2-week intervals (except the last occasion being one week further). The data was based on two treatments (A, B) of 30 calves, the alternative treatments being to counteract the debilitating effects on growth of roundworm larvae, found in feces of infected cattle, which deprives

---

[4]Occasion is used as the explanatory variable in all analyses. These are, in fact, equidistantly spaced, 14 days apart, except for a final measurement occasion 17 days after the penultimate occasion. Unlike the height data, there was no variation between subjects in the precise times of measurement within occasion. As the occasions are close in time, a more appropriate model would allow for autocorrelation of errors over time.

TABLE 11.6

Height Data for Females; 1% and 5% Observations Perturbed Upward on One Wave by 2.00 SDs

		Original			Final	
1% Observations Perturbed						
Model						
Wave[a]	Age (Linear)	Age (Quadratic $*10^{-4}$)	Wave 2 Dummy Variable	Age (Linear)	Age (Quadratic $*10^{-4}$)	Wave 2 Dummy Variable
1	0.013	-0.59	0.97	0.022	-1.36	0.82
	(0.006)	(0.55)	(0.13)	(0.004)	(0.41)	(0.10)
2	0.018	-0.97	1.19	0.022	-1.35	0.80
	(0.006)	(0.55)	(0.13)	(0.004)	(0.41)	(0.10)
3	0.020	-1.21	1.02	0.022	-1.34	0.80
	(0.006)	(0.53)	(0.13)	(0.004)	(0.41)	(0.10)
4	0.018	-0.76	1.05	0.022	-1.35	0.83
	(0.006)	(0.55)	(0.13)	(0.004)	(0.41)	(0.10)
No error	0.019	-1.02	1.03	0.022	-1.34	0.80
	(0.006)	(0.52)	(0.13)	(0.004)	(0.41)	(0.10)
5% Observations Perturbed						
Model		Original			Final	
Wave[a]	Age (Linear)	Age (Quadratic $*10^{-4}$)	Wave 2 Dummy Variable	Age (Linear)	Age (Quadratic $*10^{-4}$)	Wave 2 Dummy Variable
1	-0.016	+1.56	0.67	0.021	-1.42	0.87
	(0.007)	(0.65)	(0.16)	(0.005)	(0.46)	(0.11)
2	0.018	-0.97	1.68	0.022	-1.30	0.80
	(0.007)	(0.63)	(0.16)	(0.005)	(0.44)	(0.10)
3	0.041	-3.25	0.72	0.022	-1.42	0.80
	(0.007)	(0.61)	(0.15)	(0.005)	(0.44)	(0.11)
4	0.016	-0.15	1.05	0.024	-1.51	0.86
	(0.007)	(0.60)	(0.15)	(0.005)	(0.43)	(0.10)
No error	0.019	-1.02	1.03	0.022	-1.34	0.80
	(0.006)	(0.52)	(0.13)	(0.004)	(0.40)	(0.10)

[a] wave on which perturbed observations are added

Note. Standard errors in parentheses.

the animal of nutrients and lowers resistance to other diseases, impeding growth. Interest lies in comparing the two treatments over time. The models suggested by Kenward (1987) allowed for local variations in the rate of change of weight over time in both groups by allowing correlations of errors only within a specified range of occasions (an antedependence structure). However, we fit standard polynomial growth models with errors assumed independent across occasions, modeling growth by including terms for occasion up to the quintic and allowing these to interact with the dummy variable indicating treatment group (A vs. B). In practice, only terms up to and including a cubic were found to interact in this way. We make no claim here that polynomial models have comparable value to the locally fitted models used in Kenward's analysis. Rather, this data was used as an illustration of the effects of the iterative residual detection and deletion method in the context of polynomial growth models on the estimate of the difference between treatment groups at one point in time (occasion 5) and a comparison of this to the effect of unnecessarily simplifying the model.

Four models were fitted to this data involving all combinations of "full" and "restricted" fixed part and "simple" and "complex" random part. The "full fixed" model included all polynomial terms in occasion up to the quintic term and interactions of treatment with terms up to and including the cubic (higher powers were tested in each case and found not to reach statistical significance). The "restricted fixed" model restricted terms to the linear and quadratic terms and interactions of these with treatment group. The "complex random" model included a random coefficient for the linear term in occasion and allowed for the error variance to vary by occasion according to the square of the occasion (after previous inspection of residuals from a simple model). The "simple random" model restricted the error variance for the linear term in occasion at the individual level to be zero and restricted the error variance within the individual to be constant across occasions.

The basic model was the full fixed, complex random model and was first fitted to the data, and the effect on all parameters in the model of iterative deletion of large residuals is shown. Then, comparisons were made of the coefficient of the treatment contrast (A vs. B), designated by "Group" across the four possible models, both before and after the iterative detection and deletion of large residuals. The cutoff for deletion of large residuals was set at 3 standard deviations throughout.

Table 11.7 shows the basic polynomial model fitted to this data set. The data was centered on occasion 5 so that the estimate of the effect of treatment contrast was for this occasion.

The iterative residual detection and deletion method converged in all cases in 2 iterations. The effect of iterative deletion of large standardized

TABLE 11.7

Calf Data: Basic Polynomial Model Showing Estimates and Standard
Errors (in Brackets)

Random Between Individual	Original Model		After Residual Deletion	
Constant*constant	106.10	(21.00)	111.00	(21.70)
Occasion*constant	-1.38	(2.57)	-0.97	(2.51)
Occasion*occasion	3.10	(0.62)	2.89	(0.57)
Within individual;				
Constant*constant	15.3	(2.1)	13.5	(1.8)
Occasion^2*occasion^2	0.389	(0.062)	0.315	(0.053)
Fixed				
Constant	191.70	(5.20)	192.70	(5.10)
Treatment group (A vs. B)	3.57	(3.42)	3.64	(3.45)
Occasion; linear	16.50	(0.91)	16.27	(0.86)
Occasion; quadratic	-1.55	(0.13)	-1.54	(0.12)
Occasion; cubic ($*10^{-2}$)	-4.53	(3.87)	-3.97	(3.57)
Occasion; quartic ($*10^{-2}$)	7.69	(6.82)	7.60	(6.37)
Occasion; quintic ($*10^{-2}$)	-1.05	(1.42)	-1.01	(1.32)
Group*occasion; linear	-1.17	(0.56)	-1.02	(0.53)
Group*occasion; quadratic	-0.24	(0.06)	-0.24	(0.06)
Group*occasion; cubic ($*10^{-2}$)	6.04	(1.81)	4.24	(1.70)

TABLE 11.8

Calf Data: Comparison of Models, Estimate of Coefficient of Treatment
Group Indicator and Characteristics of Residuals

	Estimate Treatment Group [a]	of Estimate Treatment Group [b]	of SD [c]	Kurtosis [c]
Full fixed				
Simple random	3.95 (3.76)	3.92 (3.71)	7.80	1.49
Complex random	3.57 (3.42)	3.64 (3.45)	2.65	0.62
Restricted fixed				
Simple random	2.38 (3.74)	3.04 (3.73)	8.85	2.19
Complex random	2.56 (3.44)	3.13 (3.46)	6.14	1.09

[a] Estimate of treatment group indicator in original model.

[b] Estimate of treatment group indicator after residual deletion.

[c] SD and kurtosis of the residuals.

residuals is to reduce most of the coefficients in the model and their
standard errors. The small reduction in the within-individual random errors
indicated that few outliers were deleted. However, the effect of treatment
contrast, measured at the 5th occasion, increased slightly.

Table 11.8 compares the coefficients for treatment contrast at the 5th
occasion for the original model and, after residual deletion, for a range of
models that differ according to whether the fixed part was full or restricted
and according to whether the random part was complex or simple.

The difference in this coefficient before and after iterative residual
deletion was small in the models with full fixed part but, in the models
with restricted fixed part, there was substantial increase despite the fact
that terms in occasion above the quadratic — apart from the interaction
with group — failed to reach statistical significance. The deletion of large
residuals in the inappropriate restricted fixed model had the effect of moving
the coefficient of treatment group toward the value for the full fixed models.
The change to the fixed part of the model affects the size of the coefficient
of treatment group more than does the change of the random part from
simple to complex. Note that the residuals had large standard deviations
in the models with simple random components and in the restricted fixed
model, an indication that these models are not fitting the data adequately.
Moreover, the standardized kurtosis of the residuals in the simple random
models was high and over twice the value, in both full fixed and restricted
fixed models compared to the complex random models. The greater effect

of iterative residual deletion in the restricted fixed models than in the full fixed models is an indication that more large residuals are being deleted in the former case — probably due to fitting an inappropriately simple model to the data.

Although the within-occasion distribution (not shown here) generally was found to have negative kurtosis, especially at the earlier occasions, the kurtosis of the residuals was slightly positive in all models, an indication that some outliers may be present in the data. Deletion of extreme residuals would have been primarily at the later occasions and, for example, for the complex random, full fixed model, they are more than expected on the assumption of a normal distribution (1.1% vs. 0.26%). The question of model choice for this data set thus appears to be perhaps more crucial than the implementation of the procedure for detection and deletion of large residuals.

DISCUSSION

The method presented here for the iterative deletion of outliers using multilevel models has always been found to converge in few iterations for the reasonably simple models used. The effect of varying the tuning parameter, the cut-off point for the temporary deletion of residuals, is to vary the final model, the differences between the original and final models being larger when the tuning parameter is lower.

On data comprising repeated measures on height, this method is able to retrieve parameter estimates of a model fitted to data using the value of the tuning parameter in the middle of the range examined when varying proportions of the data are perturbed on each occasion in turn and by varying amounts, particularly when the proportion of observations so perturbed does not exceed 1%.

Analysis of another data set (on growth in calves) suggests that the choice of fitted model is crucial, determining the extent of change of parameter estimate with iteration, although this method appears to offer some protection from the biasing effect on key parameters of an overly simplistic model. In this data, the fixed part was found to best be modeled by polynomials that involve at least a cubic term with the random part heteroscedastic, the error variance increasing with occasion, and with the coefficient for the linear growth term varying randomly by the individual. For the best model (full fixed, complex random), little change in the parameters, and in particular of the estimate of the effect of treatment group at occasion 5, is found as a result of the iterative deletion of large residuals. For the worst model (restricted fixed, simple random), a substantial change in these estimates is found. Thus, the change in parameter estimates over the iteration process provides information on the adequacy of the original model fitted as well as on the presence or otherwise of genuine outliers. Information on the distributional characteristics of the residuals provides additional useful information on the adequacy of the model, this being shown in the case of the worst model by the large standard deviation and kurtosis of residuals.

We now examine the decision on the value for the tuning parameter, discuss possible alterations and improvements to the iterative process, and illustrate the importance of the choice of the original model.

A comparison of the alternative values of the tuning parameter for models for the height data has shown some sensitivity of parameters of the final model to the value chosen. Choosing values that are too low artificially and unnecessarily eliminates much of the true variation in the data unless a substantial portion of the data are outliers. Yet, if too large a value is chosen, real outliers are likely to contaminate the model fitted. One solution to the choice of cut-off value would be to chose it on empirical grounds through examination of the distribution of the residuals from the original model. This could be done based on a plot of the proportion of residuals at or above the to-be-set value against the corresponding estimated proportion from a normal distribution, a cut-off point being chosen when the ratio of the former to the latter shows a marked increase. The appropriate choice of this value may result from further experience of the use of this procedure.

The iterative process may be improved by, rather than deleting supposed outlying observations, replacing, these with estimates from the model, ideally with the incorporation of a random error set through the estimate of the residual standard errors at each level in the model (Goldstein, 1995). This is in the same spirit as the model-based imputation procedures of Rubin (1987) and has the advantage of allowing subsequent analyses, and in particular cross-sectional analyses, to be on the complete set of cases, thus avoiding any unrepresentativeness due to particular characteristics of outliers deleted.

Further improvements to the iterative method could be: 1. To differentially weight observations according to their residuals from the current model[5] (instead of deleting those above the cut-off value).
2. To consider residuals at more than one level in combination (at present, only residuals at the lowest level are considered for reasons given earlier). This is possible through a nested iterative procedure that starts with the deletion of large residuals at the lowest level, then cycles up the hierarchy of levels, eliminating large residuals at each of these levels in turn. At each stage, residuals at lower levels would be

---

[5]This method can be traced back to Glaisher (1872), who considered an iterative procedure for estimating weights using Bayesian principles.

reestimated and eliminated when above the specified cut-off value. An alternative is to use a dummy variable to represent any unit at higher levels with a particularly high residual, thus retaining them in the data set but eliminating their influence on the parameter estimates in the model.

3. To explicitly consider the influence (see Chatterjee & Hadi, 1986) of the data points on the model parameter estimates, possibly in addition to their residuals, as a criterion for deletion. This would give priority to deletion of observations with high residuals at extreme occasions[6] .

There is clearly scope for substantial complexity in the models used, both in the fixed and random parts. However, when the number of waves or occasions of measurement of the data is small (4 for the height data) models with complex random components are sometimes found to result in lack of convergence[7] (sometimes through oscillation between alternative estimates), although it is possible for the height data to include a random component at the individual level for the difference between the self-reported and the measured height. Such a random component was found for females but not for males, this being reflected in heterogeneous variance between waves. For the calf data (with 11 measurement occasions), it was possible to allow the linear growth parameter to be random at the individual level (thus extending the analysis of Kenward) and to allow for polynomial terms of high degree (up to quintic), and in addition, to allow for heterogeneous (increasing) variance across occasions.

Fitting too constrained a model runs the risk of eliminating, as a supposed outlier, valid values on individuals with unusual patterns of growth. For example, an observation on an early wave for an individual who is a late developer may be detected as an outlier in an inappropriate model in which the linear growth parameter is constrained to equality across individuals. We would suggest that such models be as complex as are supported by the data and result in reasonable convergence behavior over the iterative process. The distribution of residuals (both nonnormality and variance) and their change with increases in model complexity are indeed indications of the adequacy of the model fitted (assuming the true errors are normally distributed). Such an iterative procedure should in

---

[6]Note that outliers may not be influential. Moreover, influential observations may not be outliers, as with the observation having the maximum x value in a linear regression, which has high influence on the regression coefficient even if the value of y is as predicted!

[7]Although models on height data that allowed for random slope at the individual level were found for males but not for females, giving a coefficient of 0.00091(0.00019). After iterative deletion of residuals with the 3 standard deviation cutoff, the random component for slope was no longer statistically significant (coefficient 0.000037, SE 0.000071), suggesting that in this case, the random slope was an artifact of the presence of outliers in the original data.

no cases be used to justify fitting an overly simple model to the data —
rather, it may suggest particular ways in which the model could be extended
to accommodate particular subsets of observations that are generated by
different mechanisms (in Hawkins's sense).

The models used can easily be extended to include units at a higher
level than at the individual. In particular, allowing for random variation
by interviewer at a level above the individual allows the identification of
"rogue" interviewers through extreme residuals at this level. Interviewer
bias is estimated by interviewer posterior means, possibly adjusted for
socioeconomic variables, and can be modeled as varying over the position
in the sequence of interviews, allowing for systematic interviewer drift over
time. Although devised to be appropriate, particularly to longitudinal
data, this method, appropriately modified, can be used with data of other
types, for example cross-sectional hierarchical data sets. In these cases,
the iterative method can easily be adapted to screen observations at higher
levels as a priority, as when interest is in the identification of outlying units
at these levels (for example, schools or interviewers).

Further examination of this method of iterative detection and deletion of
large residuals is desirable using data with varying degrees of missingness at
particular waves, varying attrition, and varying degrees of change over time
and of model complexity. However, this investigation has suggested that
on two data sets, changes in parameter estimates result, and that a model
fitted to original data can be effectively retrieved when a particular pattern
of perturbation is added to the data. Although neither of the data sets used
here are psychological in nature, the methods here should apply to data in
the form of scales of measurement constructed in order to be normally
distributed, or that can be rendered so by a suitable transformation.

REFERENCES

Barnett, V., & Lewis, T. (1994). *Outliers in statistical data* (3rd ed.).
    Chichester: Wiley.
Diggle, P. (1998). Dealing with missing data in longitudinal studies. In G.
    Dunn & B. J. Everitt (Ed.), *Recent Advances in the Statistical Analysis
    of Medical Data* . London: Arnold.
Diggle, P., & Kenward, M. G. (1994). Informative dropout in longitudinal
    data analysis (with discussion). *Applied Statistics, 43* , 49–93.
Dixon, W. (1950). Analysis of extreme values. *Annals of Mathematical
    Statistics, 21*, 488–506.
Ecob, R., & Jamieson, B. (1992). A multilevel analysis of interviewer effects
    on a health survey. In A. Westlake (Ed.), *Survey and statistical com-
    puting* (pp. 255–268). Amsterdam: Elsevier.

Glaisher, J. (1872). On the law of locality of errors of observation and on the method of least squares. *Memoirs of the Royal Astronomical Society, 39,* 75–124.

Goldstein, H. (1989). Models for multilevel response variables with an application to growth curves. In D. Bock (Ed.), *Multilevel analysis of educational data* . New York: Academic Press.

Goldstein, H. (1995). *Multilevel statistical models* . (2nd ed.). London: Arnold.

Hawkins, D. (1980). *Identification of outliers.* London: Chapman & Hall.

Hilden-Milton, J. (1995). *Multilevel diagnostics for mixed and hierarchicial linear models.* Los Angeles: University of California.

Hodges, J. (1998). Some algebra and geometry for hierarchical models, applied to diagnostics. *Journal of the Royal Statistical Society, 60* (3, Series B), 497–536.

Hoeksma, J., & Van der Boek, C. J. (1993). Analyzing age dependent correlations in the study of growth and development. *Multilevel Modelling Newsletter, 5* (2), 6–8.

Kendall, M., & Stuart, A. (1969). The advanced theory of statistics. London: Griffin

Kenward, M. (1987). A method for comparing profiles of repeated measurements. *Applied Statistics, 36* (3), 296–308.

Langford, I. H., & Lewis, T. (1998). Outliers in multilevel data. *Journal of the Royal Statistical Society, 161* (2, Series A), 121–160.

Longford, N. (1993). *Random coefficient models.* Oxford, England: Clarendon.

Macintyre, S., Annandale, E., Ecob, R., Hunt, K., Jamieson, B., West, P., & Wyke, S. (1989). *The West of Scotland twenty-07 study: Health in the community.* Edinburgh: Edinburgh University Press.

Pan, H., Goldstein, H., & Di, G. (1992). A two level cross-sectional model using grafted polynomials. *Annals of Human Biology, 19* (4), 337–346.

Rasbash, J., Browne, W., Goldstein, H., Yang, M., Plewis, I., Healy, M., Woodhouse, G., Draper, D., Langford, I., & Lewis, T. (1998). *A user's guide to MLwiN* (Version 2.1a). Multilevel models project Institute of Education, University of London.

Rogosa, D., & Saner, H. (1995). Longitudinal data analysis examples with random coefficient models. *Journal of Educational and Behavioral Statistics, 20* (2), 149–170.

Royston, P., Altman, D. G. (1994). Regression using fractional polynomials of continuous covariates: Parsimonious parametric modelling (with discussion). *Applied Statistics, 43,* 429–467.

Rubin, D. (1987). *Multiple imputation for non-response in surveys* . New York: Wiley.

Sayer, A., & Willett, J. B. (1998). A cross-domain model for growth in adolescent alcohol expectancies. *Multivariate Behavioral Research, 33* (4), 509–543.

Ware, J. (1985). Linear models for the analysis of longitudinal studies. *The American Statistician, 39*, 95–101.

Acknowledgments

We would like to thank Dr. Patrick West and other members of the twenty-07 team for their provision of the twenty-07 data and for their help with particular questions concerning this data; also to Prof. Mike Kenward for the provision of the data on calf weight. This data is also available in Kenward (1987).

12

Estimating Interdependent Effects Among Multilevel Composite Variables in Psychosocial Research: An Example of the Application of Multilevel Structural Equation Modeling

Kenneth J. Rowe

Australian Council for Educational Research

The purpose of this chapter is to demonstrate data analytic techniques designed to account for the measurement, distributional, and hierarchically structured properties of data obtained in organizational psychological research. Whereas there is an abundance of literature advocating what should or should not be done in analyzing psychosocial data to address specific research questions, there is all too little that provides practical, illustrative examples. Notable exceptions for applications in structural equation modeling include the texts by Cuttance and Ecob (1987), Marcoulides and Schumacker (1996), and by Schumacker and Lomax (1996). For applications in multilevel analysis, the texts by Bryk and Raudenbush (1992); Goldstein (1987, 1995); Hox (1994); Kreft and de Leeuw (1998); and the users' guides by du Toit and du Toit (1999), Goldstein et al. (1998) and Rasbash et al. (2000a) are useful, as are the workshop manuals by du Toit, du Toit, and Cudeck (1999), and by

Rowe (1999a, 1999b, 2001). Thus, using a data set designed to explain variation in teachers' cognitive/affective constructions of their roles and perceptions of their work environments, this chapter illustrates the use of one-factor, congeneric measurement models to obtain maximally reliable composite variables, and the utility of multilevel analytic techniques in fitting regression and structural equation models to estimate the magnitude of the interdependent effects among those variables.

Following suggestions by Cuttance and Ecob (1987), Goldstein (1987, 1995), Kaplan and (1997a, 1997b), McArdle and Hamagami (1996), and Raudenbush (1995), the chapter demonstrates one approach to combining the analytic approaches of multilevel analysis and structural equation modeling. Specifically, the approach uses multivariate, multilevel analysis to "purge" variance–covariance matrices of the effects of nonindependence among variables, and then employing these matrices as input to fitting multiple-group structural equation models. Applications of this approach have been demonstrated by Rowe (1999a, 1999b, 2001), Rowe and Hill (1998), and by Rowe and Rowe (1999).

An Annotated Example

As part of a study of teacher and school effectiveness, the data presented here were obtained from 3,242 teachers, drawn from a cluster-designed, stratified sample of 145 elementary schools and secondary colleges, subsequently referred to as *Teacher Type* (TTYPE) and coded 1 for teachers in elementary schools and 0 for secondary schools. The gender composition (TSEX) consisted of 2,115 females and 1,127 males (coded 1 for female and 0 for male). For specific details of this larger study, see Hill, Holmes-Smith, and Rowe (1993); Hill and Rowe (1996, 1998); Rowe, Hill, and Holmes-Smith (1995); and Rowe and Hill (1998).

Teachers were asked to provide indications of perceptions of their work environments by responding to a 57-item School Organizational Health Questionnaire (SOHQ) developed by Hart and colleagues (Hart, 1994; Hart, Wearing, Conn, Carter, & Dingle, 2000). The items, each measured on 5-point, Likert-type, ordinal scales (i.e., strongly disagree to strongly agree), relate to 12 latent domains of interest to organizational psychologists. For the purposes of the illustrations presented here, only 5 of these domains were used, namely, morale, leadership support, peer support, goal congruence, and professional development. The definitions and item content for each of these 5 constructs are available from the author upon request. In particular, the related analyses focused on the measurement of these latent constructs as composite variables, as well as the simultaneous estimation of the magnitude of their interdependent effects.

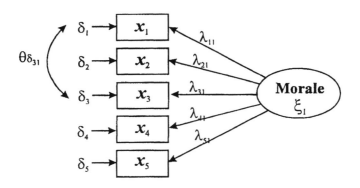

FIG. 12.1. One factor, congeneric measurement model for morale.

Construct measurement, reliability and distributional properties

Confirmatory factor analysis (CFA) was employed to compute composite scores for each of the five work environment scales using LISREL 8.30 (Jöreskog & Sörbom, 2000a) under a weighted least squares method of parameter estimation and a listwise method for deleting missing data (n = 3,033). These were obtained from fitting one-factor congeneric measurement models to the constituent ordinal-scaled, item data, based on a scaled covariance matrix (and its asymptotic estimates) of the polychoric correlations using PRELIS 2.30 (Jöreskog & Sörbom, 2000b). Composite scores computed by this method are single indices of their constituent items, each of which is weighted for its relative contribution to the composite.

Unlike traditional unit-weighted methods for computing composites, the use of factor score regression weights obtained from CFA one-factor models minimizes measurement error in the items contributing to each scale, thus increasing the reliability (and validity) of the computed scale scores. An alternative approach, using correspondence analysis, was described by Healy and Goldstein (1976). For explanatory research applications, the use of maximally reliable composite scores is crucial in fitting both single and multilevel regression models (Bryk & Raudenbush, 1992; Goldstein, 1995), as well as in fitting structural equation models (Arbuckle & Wothke, 1999; Jöreskog, Sörbom, du Toit, & du Toit, 1999; McDonald, 1985, 1994; Steiger, 1999).

The one-factor, congeneric measurement model (i.e., LISREL submodel 1) is illustrated diagrammatically in Fig. 12.1.

In matrix format, Equation 12.1 shows the regression of x_i on ξ_1 where the elements λ_{xi} are the partial regression coefficients of ξ_1 in the regression of x_i on ξ_1, namely:

$$\mathbf{x}_i = \lambda_{xi}\xi_1 + \delta_i \qquad (12.1)$$

or,

$$
\begin{bmatrix} x_1 \\ x_2 \\ x_3 \\ x_4 \\ x_5 \end{bmatrix}
=
\begin{bmatrix} \lambda_{11} \\ \lambda_{21} \\ \lambda_{31} \\ \lambda_{41} \\ \lambda_{51} \end{bmatrix}
[\xi_1] +
\begin{bmatrix} \delta_1 \\ \delta_2 \\ \delta_3 \\ \delta_4 \\ \delta_5 \end{bmatrix}.
\qquad (12.2)
$$

The assumed model implies that the covariance matrix of the observed indicators (x_i) is of the form:

$$\Sigma = \lambda_{xi}\lambda'_{xi} + \Theta_\delta \qquad (12.3)$$

where Θ_δ is a diagonal matrix with elements $\theta_{\delta i}$ indicating the variances of δ_i (i = 1, 2, 3, 4, 5).

From the parameters of Equation 12.3 the reliability (r_c) of a composite (ξ_c) is given as

$$r_c = \frac{\mathbf{w}'_c \left(\hat{\Sigma} - \hat{\Theta}_\delta \right) \mathbf{w}_c}{w'_c \hat{\Sigma} w_c}, \qquad (12.4)$$

where $\mathbf{w}_c$ is a vector of factor score (FS) regression weights that maximize the reliability of the composite. Factor score regression coefficients (FS) represent the estimated bivariate regression of the factor (ξ) on all the observed indicator variables, given by $FS = \Lambda'\Sigma^{-1}$, where Λ is the estimated factor pattern matrix and Σ is the estimated covariance matrix of the observed indicators (see Jöreskog & Sörbom, 1989, p. 93; Lawley & Maxwell, 1971, p. 109). Factor score estimates (ξ_i) may be computed for any individual i with observed scores x_i, using the simple product: $\xi_i = FS\ \mathbf{x}_i.$

For specific details of these well established but all too rarely used procedures, see Alwin and Jackson (1980), Fleishman and Benson (1987), Jöreskog (1971, 1990, 1994), Werts, Rock, Linn, and Jöreskog (1978). Further details including the rationale for this approach to computing composite variables and their reliabilities have more recently been outlined and demonstrated by Holmes-Smith and Rowe (1994), McDonald (1996), Rowe and Hill (1998), and by Rowe and Rowe (1999). For the present case, the relevant PRELIS and LISREL files for estimating the parameters of the composite variable morale are available from the author upon request.

From Equations 12.1 and 12.3 the item factor score regression coefficients for each composite, and the reliability coefficients for the composite scores from Equation 12.4 were computed and are presented in Table 12.1. For comparative purposes, the traditional lower bound estimate of reliability for each composite, namely, Cronbach's (1951) standardized item alpha (α), is given in the final column of Table 12.1.

A proportionally weighted scale score for the composite variable morale that takes into account the individual and joint measurement error of the indicators mo_1, mo_2, mo_3, mo_4, and mo_5, was computed as a continuous variable for each case as follows (e.g., using SPSS format nomenclature):

compute **MORALE** =
$(mo_1*0.127)+(mo_2*0.186)+(mo_3*0.185)+(mo_4*0.299)+(mo_5*0.203)$,

where mo_1, mo_2, mo_3, mo_4, and mo_5 are the raw score ratings made by each respondent on the five indicator items, respectively. This process ensures that the estimation of the scale/composite variable morale (adjusted for measurement error) is proportionally weighted by the actual contribution made by each indicator. Note that these proportionally weighted FS regression coefficients add to 1; hence, the scale/composite score will range from a minimum of 1 to a maximum of 5. This means that the composite variable morale (ξ_{c1}) and the associated scales (ξ_{c2}, ξ_{c3}, ξ_{c4}, ξ_{c5}), as given in Table 12.1, have the advantage of all being measured in the same metric. The computed scale/composite score for morale (and similarly in the case for each construct) may now be used in explanatory analyses, namely, in fitting structural equation models and multilevel models.

Before doing so, however, it is important to examine the distributional properties of the continuous variables to be used in subsequent explanatory modeling. For such purposes, PRELIS gives a detailed summary of the descriptive parameters (i.e., first-, second-, and third-order moments) and provides both univariate and multivariate tests of zero skewness and zero kurtosis. For the present case, the computed estimates are summarized in the upper and middle sections of Table 12.2.

Because the relevant estimates (as given in the middle section of Table 12.2) indicated that all five variables were significantly nonnormal, the raw composite scale scores were recomputed as normal scores. This can be easily and efficiently done using MLn or MLwiN (Rasbash, Woodhouse, Yang, & Goldstein, 1996, Rasbash, Browne, Healy, Cameron, & Charlton, 2000b), which use a method of rescoring (via the NSCOR command) that assigns expected values from the standard normal distribution according to the ranks of the original scores in the form of normal equivalent deviates (NED). The obtained normalized estimates given in the lower portion of Table 12.2 are the ones used for explanatory modeling here.

TABLE 12.1
Composite Scale Parameters

Scale	N^a	TD^b	Item Weights^c					r_c	$\hat{\lambda}_c$	$\hat{\Theta}_c$	α^d
			mo1	mo2	mo3	mo4	mo5				
ξ_{c1} - Morale			.140	.206	.205	.330	.224				.884
	3,173	3,1	.127	.186	.185	.299	.203	.902	.942	.096	
			ls1	ls2	ls3	ls4	ls5				
ξ_{c2} - Leadership			.149	.069	.367	.285	.216				.880
support	3,162	-	.137	.064	.338	.262	.199	.930	.948	.068	
			ps1	ps2	ps3	ps4	ps5				
ξ_{c3} - Peer Support		4,1	.133	.246	.118	.174	.448				.845
	3181	5,1	.119	.220	.105	.155	.401	.900	.942	.099	
			gc1	gc2	gc3	gc4	gc5				
ξ_{c4} - Goal		3,2	.296	.096	.174	.286	.303				.815
congruence	3,174	4,1	.256	.083	.151	.248	.262	.871	.924	.126	
			pd1	pd2	pd3	pd4	pd5				
ξ_{c5} - Professional		5,2	.312	.374	.179	.121	.216				.801
development	3,170	2,1	.259	.311	.149	.101	.180	.890	.940	.109	

Note. The r_c is the composite scale reliability coefficient calculated from the maximally weighted factor score regression coefficients obtained from fitting one-factor congeneric measurement models to the constituent scale items, given by:

$$r_c = \frac{w'_c \left(\hat{\Sigma} - \hat{\Theta}_\delta \right) w_c}{w'_c \hat{\Sigma} w_c},$$

where $\mathbf{w_c}$ is the vector of factor score regression weights. $\hat{\Lambda}_c = \sigma_c \sqrt{r_c}$ is the estimate of that part of the variance in the vector of indicator variables ($\underline{x}$) that is explained by the latent or composite variable ξ_c. The standard deviation estimates (σ_c) are given in Table 12.2 (normal scores). $\hat{\Theta}_c = \sigma_c^2 (1 - r_c)$ is the estimate of the remaining variance in the indicator variables not explained by the composite variable - i.e., measurement error variance. The variance variance estimates (σ_c^2) are the diagonals of the variance-covariance matrix given in Table 12.2 (Normal Scores).

[a] N= the number of cases with complete data.

[b] TD(Θ_δ)= indicates correlated error variance estimates, computed on substantive grounds.

[c] The second row for each scale shows the proportionally weighted FS regression coefficients.

[d] Cronbach's standardized item alpha.

TABLE 12.2

Descriptive Estimates for Five Work Enviornment Scales ($N = 3{,}033$)

Means, Standard Deviations, Variances[a] –Covariances[b], and Correlations[c] (Raw Scores):

	Mean	SD	Morale	Leadsup	Peersup	Goalcon	Profdev
Morale	3.356	0.847	**0.7180**	*0.669*	*0.684*	*0.766*	*0.562*
Leadsup	3.478	0.942	0.5337	**0.8878**	*0.581*	*0.655*	*0.568*
Peersup	3.619	0.761	0.4410	0.4162	**0.5786**	*0.635*	*0.579*
Goalcon	3.573	0.727	0.4722	0.4491	0.3511	**0.5289**	*0.555*
Profdev	3.212	0.838	0.3390	0.4488	0.3693	0.3383	**0.7021**

Univariate and Multivariate Tests of Normality (Raw Scores):

	Skewness			Kurtosis			Sk. & Kurtosis	
	Est.	z-score	p	Est.	z-score	p	X^2	p-value
Morale	-0.379	-4.426	<.001	-0.260	-2.903	0.002	28.02	<.001
Leadsup	-0.501	-4.943	<.001	-0.211	-2.349	0.009	29.95	<.001
Peersup	-0.436	-4.687	<.001	0.006	0.092	0.463	21.98	<.001
Goalcon	-0.319	-4.115	<.001	0.037	0.441	0.329	17.13	<.001
Profdev	-0.199	-3.264	0.001	-0.393	-4.404	<.001	30.07	<.001
Multivariate		19.020	<.001		17.290	<.001	660.7	<.001

Means, Standard Deviations, Variances[a] -Covariances[b] and Correlations[c] (Normal Scores):

	Mean	SD	Morale	Leadsup	Peersup	Goalcon	Profdev
Morale	-0.002	0.992	**0.9836**	*0.678*	*0.700*	*0.773*	*0.574*
Leadsup	-0.004	0.983	0.6611	**0.9662**	*0.604*	*0.662*	*0.578*
Peersup	-0.002	0.993	0.6878	0.5891	**0.9858**	*0.647*	*0.586*
Goalcon	-0.004	0.990	0.7590	0.6434	0.6354	**0.9792**	*0.561*
Profdev	-0.001	0.996	0.5645	0.5663	0.5794	0.5527	**0.9923**

[a] Variance estimates, in bold type, on the diagonal.

[b] Covariance estimates, in normal type, below the diagonal.

[c] Correlations, in italics, above the diagonal.

EXPLAINING VARIATION

To explain variation in teachers' morale scores, several models are fitted to the data. For the purpose of evaluating the obtained parameter estimates, three models are fitted, namely; (a) multilevel variance components models for each of the five composite constructs using iterative generalized least squares estimation (IGLS, which is maximum likelihood under normality; see Goldstein, 1986; Goldstein & Rasbash, 1992); (b) a multilevel regression model, under IGLS with morale as the response variable and the related constructs as explanatory variables; and (c) a two-level structural equation model using maximum likelihood estimation on the separate teacher-level and school-level covariance matrices, simultaneously. To assist presentation and interpretation of the results, specifications for the models and their solutions are given first, followed by comments.

MODEL 1: FITTING SIMPLE, TWO-LEVEL, VARIANCE COMPONENTS MODELS

An initial step in any explanatory modeling of data is to determine the proportion of variance in both response and explanatory variables that may be due to the data structure. In the present case, we have data for 3,242 teachers clustered in 145 primary and secondary schools. Under such circumstances, the fitting of single-level regression models, or structural equation models of the LISREL kind, are inappropriate because such models assume that there is random variation among teachers, and that sample variables, regardless of level, are normal and independently distributed (NID). That is, because these models assume single-level data, their use can only be justified if the intraclass (intragroup) correlation estimate (ρ - rho) is negligible. If ρ is substantially different from zero, fitting any single-level model not only violates the assumptions of independence, but gives rise to problems affecting validity of statistical conclusions, such as misestimated parameters and their standard errors, and an increased likelihood of generating Type I errors.

Thus, to determine the proportion of variance in each of the teacher work environment scales due to between-school differences, simple two-level variance components (VC) models are fitted. Using the subscript i to refer to the teacher and the subscript j for the school, this model may be written as:

$$y_{ij} = \beta_{0j}(\mathbf{X}_0) + (u_{0j} + e_{ij}), \qquad (12.5)$$

where, y_{ij} is the morale score (for example) for teacher i in school j, β_{0j} is the mean morale score for teachers in the sample of schools (intercept). The $(\mathbf{X}_0)$ term is a vector of unities that operate as indicators to define the data structure, u_{0j} is a residual that varies randomly between schools, and e_{ij} is a random variable (assumed to have a mean of zero) representing the sum of all other influences on y_{ij}. The term β_{0j} constitutes the fixed part of the model; u_{0j} and e_{ij} form the random part of the model. The distribution assumptions for the random coefficients are:

$u_{0j} \sim NID(0, \sigma_{u0}{}^2)$, where $\sigma_{u0}{}^2$ is the variance of the level-2 (school) residuals u_{0j},

$e_{ij} \sim NID(0, \sigma_e{}^2)$, that is, $\sigma_e{}^2$ is the variance of the level-1 (teacher) residuals e_{ij},

u_{0j} and e_{ij} are normal and independent (NID),

and the intraschool correlation is given by $\rho = \sigma_{u0}{}^2/(\sigma_{u0}{}^2 + \sigma_e{}^2)$. This correlation provides an estimate of the proportion of the total variance in teachers' morale scores that is due to variation between schools. To determine the extent to which schools differ in their mean levels of morale, the ratio of the $\hat{\sigma}_{u0}{}^2$ estimate to its standard error $[\hat{se}(\hat{\sigma}_{u0}{}^2)]$ can be referred to the usual Gaussian distribution (t-value). The upper section of Table 12.3 summarizes the results of fitting VC models to each of the variables of interest and the lower section of Table 12.3 provides the same estimates adjusted for the effect of teacher type (TTYPE) and teacher gender (TSEX). To fit these models, we used the multilevel analysis program MLwiN (Rasbash et al., 2000b). It should be noted that previous MANOVA analyses of these data indicated that the interaction effect of TTYPE and TSEX was not significant (Rowe, 1995).

COMMENT

From Table 12.3, the significant fixed effect of TTYPE on each of the composite variables indicates that teachers in elementary schools have more positive perceptions of their work environments than their counterparts in secondary schools. Of special interest from the random parts of the models is that the proportion of between-school residual variance for each of the variables is statistically significant. That is, the ratio of the parameter estimates to their respective standard errors are all greater than 1.96 (i.e., the critical t-value under the normal distribution at the $p < 0.05$ α level), indicating that teachers' perceptions of their work environments are significantly influenced by the contexts (schools) in which they work. Again, it is important to stress that under such circumstances, multilevel analyses are essential if correct statistical and substantive conclusions are to be drawn from any subsequent explanatory modeling procedures.

TABLE 12.3

Variance Components for Five Variables Showing Proportions of
Between-Schools and Within-Teachers Residual Variance: Parameter Estimates
and Standard Errors in Parentheses (3,033 Teachers in 145 Schools)[a]

Base Variance Components Models

Response Variables	Fixed Sch. Intercepts		Random Residual Variance						Total
			Between-Schools			Within Teachers			
	β_{0j}	(SE)	σ_{u0}^2	(SE)	%	σ_e^2	(SE)	%	σ_T^2
Morale	0.179	(.052)	0.336	(.046)	34.2	0.647	(.017)	65.8	0.983
Leadsup	0.154	(.047)	0.259	(.037)	25.8	0.743	(.020)	74.2	1.002
Peersup	0.134	(.045)	0.239	(.035)	23.3	0.787	(.021)	76.7	1.026
Goalcon	0.167	(0.46)	0.251	(.036)	25.6	0.730	(.019)	74.4	0.981
Profdev	0.168	(.045)	0.232	(.034)	23.4	0.761	(.020)	76.6	0.993

Variable	Fixed					
	Sch. Intcpt		TType		TSex	
	β_{0j}	(SE)	β_1	(SE)	β_2	(SE)
Morale[b]	-.17	(.070)	0.60	(.079)	-0.07	(.033)
Leadsup	-.10	(.068)	0.48	(.041)	-0.09	(.035)
Peersup	-.23	(.064)	0.49	(.034)	0.06	(.036)*
Goalcon	-.20	(.062)	0.55	(.031)	0.07	(.035)*
Profdev	-.27	(.057)	0.59	(.037)	0.07	(.035)*

Variable	Random (Residual Variance)					
	Between Schools			Within Teachers		
	σ_{u0}^2	(SE)	%	σ_e^2	(SE)	%
Morale[b]	0.23	(.033)	26.3	0.65	(.017)	73.7
Leadsup	0.20	(.029)	21.1	0.74	(.020)	78.9
Peersup	0.16	(.025)	16.8	0.79	(.021)	83.2
Goalcon	0.15	(.024)	17.0	0.73	(.019)	83.0
Profdev	0.12	(.020)	13.3	0.76	(.020)	86.7

[a] Sample size with complete data and ≥ 5 teachers within each school.

[b] The log likelihood deviance estimate [-2*log(lh)] for the fitted model for morale = 7609.9.

* Not statistically significant at the $p < 0.05$ level by univariate two-tailed test.

MODEL 2: FITTING A CONDITIONAL, MULTILEVEL REGRESSION MODEL

To estimate the proportion of variance in morale due to the effects of the related variables of interest and allowing for the structure of the data, a conditional multilevel regression model was fitted to the total teacher data. For this model, morale (Y_{ij}) for teacher i in school j was used as the response variable, and TSEX (X_{1ij}), TTYPE (X_{2ij}), leadsup (X_{3ij}), peersup (X_{4ij}), goalcon (X_{5ij}), and profdev (X_{6ij}) were fitted as level-1 explanatory variables. For school j, a linear relationship between these variables can be written as:

$$
\begin{aligned}
Y_{ij} &= \beta_{0j}(X_0) + \beta_{1j}X_{1ij} + \beta_{2j}X_{2ij} + \beta_{3j}X_{3ij} \\
&+ \beta_{4j}X_{4ij} + \beta_{5j}X_{5ij} + \beta_{6j}X_{6ij} + u_{0j} + e_{ij}
\end{aligned} \tag{12.6}
$$

It should be noted that estimation of the variance in Y_{ij} given by Equation 12.6 is conditional on the linear combination of the fitted explanatory variables $(X\mathrm{s})$. Suppose, say, that the scale for each of X_3, X_4, X_5, and X_6 is chosen so that 0 (zero) represents average leadsup, peersup, and so forth. The intercept β_{0j} in this within-school relationship is the average change in morale (Y_{ij}) for each unit of change in the six explanatory variables $(X_1, X_2, X_3, X_4, X_5, \text{ and } X_6)$ jointly. For notational consistency, the intercept is written as X_0 $(= 1)$. The residual term u_{0j} varies randomly between schools and e_{ij} is also a random variable — assumed to have a mean of zero — that represents the sum of all level-1 (teacher) influences on Y_{ij}, other than those of the fitted explanatory variables. Because β_{0j}, β_{1j}, β_{2j}, β_{3j}, β_{4j}, β_{5j}, and β_{6j} in general can and do vary across schools, these coefficients are treated as random variables at level 2 (school).

For illustrative purposes, say we also wish to estimate the effect of school average goal congruence, over and above its effect at the individual teacher level goalcon (X_{5ij}). We can do this by fitting avgoal (Z_{1j}) as a contextual variable where each teacher within a given school is assigned their respective within-school mean score for goal congruence. Thus, a between-school model for β_{0j} and β_{1j} (say) in terms of Z_{1j} can be written as:

$$
\beta_{0j} = \beta_0 + \beta_{01}Z_{1j} + u_{0j}. \tag{12.7}
$$

The coefficient β_0 is the mean morale for average scores on X_3, X_4, X_5 and X_6 in schools for which Z_{1j} (avgoal) $= 0$, and β_{01} is the average effect of Z_{1j} on mean morale for teachers experiencing average leadsup (X_{3ij}), peersup (X_{4ij}), goalcon (X_{5ij}), and profdev (X_{6ij}). In general, γ_{01} represents the rate of change of the group mean of the response variable

Y_{ij} with Z_{1j}. Likewise, β_{01} is the average effect of X_3, X_4, X_5, and X_6 on morale (Y_{ij}) in schools with high avgoal (Z_{1j}), and γ_{11} is the average increment to this slope attributable to differences in Z_{1j}. The random variables u_{0j} and u_{1j} represent the influences on the β's not accounted for by Z_{1j}. Note, however, that unless our research questions specify such analyses, we are not required to model both β_{0j} and β_{1j} as functions of Z_{1j}. For the purposes of this chapter, the simpler two-level model is fitted:

$$\begin{aligned} Y_{ij} &= \beta_{0j}(X_0) + \beta_{1j}X_{1ij} + \beta_{2j}X_{2ij} + \beta_{3j}X_{3ij} \\ &+ \beta_{4j}X_{4ij} + \beta_{5j}X_{5ij} + \beta_{6j}X_{6ij} + \gamma_{01}Z_{1j} + u_{0j} + e_{ij}. \end{aligned} \quad (12.8)$$

From this model, we are interested in estimating the magnitude of the effect of school average goal congruence, (avgoal, Z_{1j}), over and above that which operates at the individual teacher level, namely, goalcon (X_{5ij}).

The parameter estimates (β's, γ, u_{0j}, and e_{ij}) for the solution to the multilevel regression model specified by Equation 12.8, under iterative generalized least squares (IGLS) estimation, are given in Table 12.4. For comparative purposes, the ordinary least squares (OLS) estimates are also provided. It is important to note, however, that the OLS solution to the teacher-level regression model given in Table 12.4, like all single-level outcomes of fitting the general linear model, provides only for random variation between teachers (level-1 units). That is, the grouping of teachers into schools is ignored (it is assumed that u_{0j} does not exist, or that its variance $\sigma_{u0}^2 = 0$) and the data are treated as a single sample of Σn_i observations on the Y, X, and Z variables (see Draper & Smith, 1981; Mardia, Kent, & Bibby, 1979; Rowe, 1989).

Comment

The results presented Table 12.4 indicate that the seven explanatory variables accounted for approximately 71% of the variance in morale. The log-likelihood statistic for the model was -2log(lh) = 4,568.0, indicating a significant reduction in deviance units ($p < 0.0001$) from the base variance components model given in Table 12.3 (i.e., 7,609.9) due to the fitted variables. Peer support (peersup) and goal congruence (goalcon) constituted the major explanatory variables at the teacher level, and school average goal congruence (avgoal) had a large and significant effect on accounting for variation in morale, over and above that at the individual teacher level.

A comparison of the multilevel (IGLS) and ordinary least squares (OLS) parameter estimates given in Table 12.4 indicate similar magnitudes at the teacher level (as expected), but the OLS estimate for avgoal is artificially inflated over the IGLS estimate ($\gamma_{1OLS} = 0.326$, cf. $\gamma_{1IGLS} = 0.249$), and

TABLE 12.4

Variation in 3,033 Teachers' Morale Scores in 145 Schools, Showing
Unstandardized and Standardized IGLS and OLS Solutions; Fitted Estimates
With Standard Errors in Parentheses

Explanatory Variables	Multilevel Model		Single-Level Model	
	IGLS Est. (SE)	Standardized Estimate	OLS Est. (SE)	Standardized Estimate
Fixed:				
Constant ($\beta_{0j} X_0$):	0.028 (.034)	-0.099	0.061 (.019)*	-0.001
Teacher Level:				
X_1 TType (β_1)	0.060 (.044)	0.031	0.011 (.025)	0.008
X_2 TSex (β_2)	-0.098 (.020)	-0.050	-0.098 (.021)	-0.050
X_3 Leadsup (β_3)	0.187 (.015)	0.188	0.185 (.014)	0.188
X_4 Peersup (β_4)	0.253 (.014)	0.248	0.252 (.014)	0.248
X_5 Goalcon (β_5)	0.373 (.015)	0.369	0.375 (.015)	0.370
X_6 Profdev (β_6)	0.044 (.013)	0.041	0.044 (.013)	0.040
School Level:				
Z_1 Avgoal (γ_1)	0.249 (.043)	0.041	0.326 (.025)	0.182
Random:				
σ_{0u}^2 (Var of Sch. Intercepts)	0.038 (.006)			
σ_e^2 (Teacher-level Variance)	0.249 (.007)			
% of Variance explained[a]:	70.8		71.0	
-2log(lh)	4568.0			

[a] This is calculated from the difference between the total number of units of
variation for morale from the variance components model (as given in Table 12.3;
that is, $0.336 + 0.647 = 0.983$) and the sum of the residual school and teacher-level
variance, from the fitted multilevel regression model above (i.e., 0.038+ 0.249),
divided by 0.983. That is, $[(0.983-0.287)/0.983] \times 100 = 70.8$ per cent.
* Statistically significant beyond the $p < 0.05$ α level by univariate two-tailed test;
that is, the parameter estimate is greater than twice its standard error.

its standard error is almost half of that obtained from the IGLS solution (i.e., 0.025 and 0.043, respectively). It should be noted that the shortened confidence intervals for avgoal from the OLS solution could lead to the mistaken conclusion that avgoal has greater explanatory power than it actually has. In instances where the parameter estimates for such higher level variables have borderline significance from fitting single-level models, Type I errors are frequent, and the related statistical conclusion validity becomes particularly problematic (Aitkin, Anderson, & Hinde, 1981; Rowe, 1992).

STRENGTHS AND LIMITATIONS OF MULTILEVEL AND STRUCTURAL EQUATION MODELS

The results of fitting conditional multilevel regression models of the kind illustrated by Equations 12.7 through 12.12 and summarized in Table 12.4, however, do not provide information about the magnitude of the interdependent effects among the constructs. Although it is possible to include interaction terms as fixed effects in models of the kind specified by Equation 12.8, for example, researchers typically confine their modeling applications to the specification of a limited number of response variables (usually only one) as has been done so far here. Moreover, because the multilevel models fitted here are mere extensions of the general linear model ($Y = \beta X + \varepsilon$), which specifies Y (response or dependent variable) to be a simple linear sum of the effects (β) of the fitted explanatory (or independent) variables (X), they make no allowance for examining the *structure* of the covariance matrix (Σ) among the X and Y variables. Under such circumstances, it is not possible to estimate jointly the direct, indirect, and total effects operating among the X and Y variables. Nevertheless, it is possible to estimate such effects simultaneously within a structural equation modeling framework.

Fitting unconditional structural equation models (SEMs) has two key advantages. First, they provide a means of estimating the magnitude of direct, indirect, and total effects among variables jointly. Second, they can account for measurement error in both the observed and the latent variables. However, there are major limitations and disadvantages when the data are multilevel or hierarchically structured as we have here. The use of SEM approaches assumes that the sample variables, regardless of level, are independently distributed in a multivariate population. That is, unconditional SEM models assume single-level data and can only be fitted justifiably if the intraclass (intragroup) correlations (ρ — rho) are not significantly different from zero. If the ρ's are significantly different from zero, fitting any single-level model not only violates the assumptions of independence but also gives rise to several problems affecting statistical conclusion validity, including misestimated parameters and their standard errors, with important ramifications for the substantive interpretation of findings.   Short of applying procedures for purging computed variance–covariance matrices of the assumptions of independence (demonstrated here), there is no simple adjustment of the structural modeling framework that can be made in order to deal with nonindependent observations in hierarchically structured data (Cuttance, 1987, chap. 13).

Although it is possible to model a few groups separately in a multisample analysis, such an approach becomes intractable with more than a small number of groups (i.e., a maximum of 10 in LISREL 8; Jöreskog & Sörbom, 2000a). In the present case, we have data from more than 3000 teachers that are clustered within 145 schools. In the event that solutions were possible, the outputs from multiple-group analyses of this magnitude using SEM or MANOVA, for example, would be uninterpretable! Further, the crucial drawback of fitting single-level SEM models to multilevel data is that they do not allow for reliable modeling of substantive relationships across levels (e.g., school effects on teachers, or vice versa). Moreover, such models typically lack power due to the smaller number of observations in the higher level units.

Estimating the variance–covariance matrices at each level

To illustrate one approach to solving the above-mentioned problem, a multivariate, multilevel model was fitted to the five composite work environment variables to estimate the variance–covariance matrices (adjusted for TTYPE and TSEX) for the school and teacher levels. Following procedures outlined and suggested by several authors (du Toit & du Toit, 1999; Goldstein, 1995, chap. 4; McDonald & Goldstein, 1989; Muthén, 1994; Raudenbush, 1995), the purpose of this approach was to partition the variances and covariances among the five composites into separate school- and teacher-level variance–covariance matrices, as a prelude to fitting an explanatory two-level (teacher and school) structural equation model.

To define the five-variate multivariate model for our present case, schools were treated as level-3 units, individual teachers as level-2 units, and the within-teacher measurements on the five composite variables as level-1 units. It should be noted that in this case, level-1 variation is not specified because " ... level 1 exists solely to define the multivariate structure" (Goldstein, 1995, p. 70). This yielded 15,165 measurements (i.e., 3,033 × 5 variables) clustered with 3,033 teachers and 145 schools. Thus, teachers' scores on the five composite variables (y_{ijk}) may be written as in Equation 12.9 where the variables (y_i) are regarded as the level-1 structure, grouped within teachers (j — level 2), with teachers grouped within schools (k — level 3):

$$
\begin{aligned}
y_{ijk} &= (\beta_{Tjk}x_{1jk} + \beta_{Gjk}x_{2jk}) + \beta_{1jk}y_{1ijk} \\
&+ \beta_{2jk}y_{2ijk} + \beta_{3jk}y_{3ijk} + \beta_{4jk}y_{4ijk} + \beta_{5jk}y_{5ijk}, \quad (12.9)
\end{aligned}
$$

in which y_{1ijk}, y_{2ijk}, $\ldots y_{5ijk}$ are the dummy indicators (0, 1) of the scores on each of the five variables for teacher j in school k; x_{1jk} and x_{2jk} are dummy explanatory variables for TTYPE (1 = primary; 0 = secondary), and TSEX (1 = female; 0 = male), respectively; and β_{Tjk} is the coefficient for the fixed effect of TTYPE, β_{Gjk} is the corresponding coefficient for the effect of TSEX. β_1, β_2, $\ldots \beta_5$ are fixed parameters defining the adjusted means (intercepts) of the five variables over all schools.

The set of five equations for the intercepts that model variation around the means (β_1, β_2, $\ldots \beta_6$) of the five composite response variables (morale, $\ldots$ profdev) adjusted for the effects of the fixed explanatory variables TTYPE (x_{1jk}) and TSEX (x_{2jk}) may be written as:

$$
\begin{aligned}
\beta_{1jk} &= \beta_1 + v_{1k} + u_{1jk} \\
\beta_{2jk} &= \beta_2 + v_{2k} + u_{2jk} \\
\beta_{3jk} &= \beta_3 + v_{3k} + u_{3jk} \\
\beta_{4jk} &= \beta_4 + v_{4k} + u_{4jk} \\
\beta_{5jk} &= \beta_5 + v_{5k} + u_{5jk}. \quad (12.10)
\end{aligned}
$$

in which v_{1k}, v_{2k} $\ldots v_{5k}$ are the random terms required to estimate the residual variation at the school level (level 3), and u_{1jk}, u_{2jk}, $\ldots u_{5jk}$ are the random terms from which estimates of the residual variation at the teacher level (level 2) may be obtained.

Of particular interest in our present case, v_{1k}, v_{2k}, $\ldots v_{5k}$ are the residuals representing the unique contribution of the schools (k) above that explained by their means (βs), and u_{1jk}, u_{2jk} , $\ldots u_{5jk}$ are residual terms representing the unique contribution of teacher j in school k. These equations may be estimated jointly to obtain estimates of the variances and covariances among the five variables (adjusted for TTYPE and TSEX) at the school level and at the teacher level, as well as their intercorrelations. Note that the level-2 and level-3 residual variances and covariances among the five work environment variables are the between-teacher and between-school variances, respectively — purged of the effects of TTYPE and TSEX.

These variances and covariances among the residual terms at the school level (Ω_v) and at the teacher level (Ω_u) may be written (in lower triangular matrix format) for the random part of the model, as:

$$\Omega_v = \begin{matrix}
\sigma_{v11}^{2} & & & & \\
\sigma_{v21} & \sigma_{v22}^{2} & & & \\
\sigma_{v31} & \sigma_{v32} & \sigma_{v33}^{2} & & \\
\sigma_{v41} & \sigma_{v42} & \sigma_{v43} & \sigma_{v44}^{2} & \\
\sigma_{v51} & \sigma_{v52} & \sigma_{v53} & \sigma_{v54} & \sigma_{v55}^{2}
\end{matrix} \qquad (12.11)$$

$$\Omega_u = \begin{matrix}
\sigma_{u11}^{2} & & & & \\
\sigma_{u21} & \sigma_{u22}^{2} & & & \\
\sigma_{u31} & \sigma_{u32} & \sigma_{u33}^{2} & & \\
\sigma_{u41} & \sigma_{u42} & \sigma_{u43} & \sigma_{u44}^{2} & \\
\sigma_{u51} & \sigma_{u52} & \sigma_{u53} & \sigma_{u54} & \sigma_{u55}^{2}.
\end{matrix} \qquad (12.12)$$

Note again the distributional assumptions for v_{1k}, v_{2k}, $\ldots v_{5k}$ $\sim$ $N(0, \Omega_v)$, and for u_{1jk}, u_{2jk}, $\ldots u_{5jk}$ $\sim$ $N(0, \Omega_u)$. Specific details of multivariate multilevel model specifications and related applications are given by Goldstein (1995), Rasbash et al. (2000a), and Woodhouse (1996).

The results of partitioning the relevant variance–covariance and product-moment inter correlation estimates into their between-school and within-teacher components are given in Table 12.5. Specific details of the relevant MLn/MLwiN macro and output is available from the author, together with the equivalent batch file for processing by PRELIS 2.30.

Comment

The results in Table 12.5 raise several points. First, consistent with the earlier results from fitting the variance components models for each of the five composites (see Table 12.3), the findings related to the fixed parameters indicate that the overall effect of TTYPE was significant but TSEX was not. That is, independent of Teacher Gender, teachers in elementary schools indicated significantly more positive perceptions of their work environments than their counterparts in secondary schools.

Second, from the random part of the model, note that the magnitudes of the variance–covariance estimates at the teacher level are consistently

TABLE 12.5

Summary of Results From Fitting Three-Level Multivariate Model: Parameter Estimates (and Standard Errors) for Fixed Part, Lower Triangular Matrices of Residual Variance–Covariance Estimates at the School Level and at the Teacher Level (in Bold)[a] for 3,033 Teachers in 145 Schools[b]

Fixed Parameters				Estimates(SE)	
Teacher Type (TTYPE)				0.517 (.058)*	
Teacher Gender (TSEX)				0.009 (.028)	
Random:					
Composite	Morale	Leadsup	Peersup	Goalcon	Profdev
Morale	0.2307	*0.694*	*0.801*	*0.921*	*0.688*
	0.6471	***0.642***	***0.651***	***0.705***	***0.500***
Leadsup	0.1456	0.1909	*0.562*	*0.678*	*0.486*
	0.4456	**0.7443**	***0.585***	***0.625***	***0.567***
Peersup	0.1500	0.0957	0.1520	*0.811*	*0.700*
	0.4652	**0.4481**	**0.7892**	***0.583***	***0.525***
Goalcon	0.1714	0.1148	0.1225	0.1502	*0.677*
	0.4852	**0.4815**	**0.4426**	**0.7315**	***0.490***
Profdev	0.1162	0.0746	0.0959	0.0922	0.1235
	0.3515	**0.4274**	**0.4077**	**0.3661**	**0.7629**

Note: Correlation estimates (in italics) are given above the diagonal. Correlations at the teacher-level are in bold italics and those at the school level are in normal italics.

[a] Residual variance estimates are given on the diagonal and covariances below the diagonal. Teacher-level variance and covariance estimates are presented in bold type, and school-level convariances in normal type.

[b] Sample size with complete data and ≥ 5 teachers within each school.

larger than those at the school level. This finding is also commensurate with the results obtained from the fitted variance components models for each of the composite constructs given in Table 12.3. Third, the lack of systematic variation in the magnitudes of the product-moment correlation estimates at the teacher and school levels should be noted.

Fourth, the data shown in Table 12.5 raise the crucial question of what covariance matrices should be analyzed. If the research questions relate to teachers, then the teacher-level matrices should be analyzed if correct parameter estimates, standard errors, and model-fit statistics are to be obtained. A similar point is relevant if research questions are focused on relationships at the school level. Finally, and above all, the data presented in Table 12.5 underscore the point that fitting structural equation models to omnibus correlation or variance–covariance matrices that have not been adjusted for the inherent hierarchical structure of the data increases the risk of yielding misleading results at best, and meaningless results at worst.

MODEL 3: FITTING A MULTILEVEL STRUCTURAL EQUATION MODEL

From the literature related to teacher stress and quality of work life (e.g., Brenner, Sörbom, & Wallius, 1985; Cooper & Payne, 1992; Hart et al., 2000; Kyriacou & Pratt, 1985; McCormick & Solman, 1992), the proposed recursive model shown in Fig. 12.2 was tested simultaneously for fit to the data at the teacher- and school-levels.

This unconditional model posits significant direct effects on teachers' morale (morale) from leadership support (leadsup), peer support (peersup), and goal congruence (goalcon), and indirect effects from professional development (profdev), mediated by profdev, peersup, and goalcon. Moreover, there are indirect effects from goalcon on morale, mediated by peersup. In testing the data/model fit for the teacher and school levels, the LISREL method for submodel 3b can be used (see Jöreskog & Sörbom, 1989, pp.189–190). This is the most general of all covariance structure models. Its notable feature is that it contains only y (observed) and η (latent) variables. In the two-level case, the structural relations among the latent or composite variables (η) for teacher i in school j, may be written as:

$$\eta_{ij} = B_{ij}\eta_{ij} + \zeta_{ij} \qquad (12.13)$$

where B_{ij} is the matrix of effect relationships among the composite constructs (η_{ij}) at the teacher level (i) and at the school level (j); ζ_{ij} is the vector of prediction residuals in the structural equations at the two

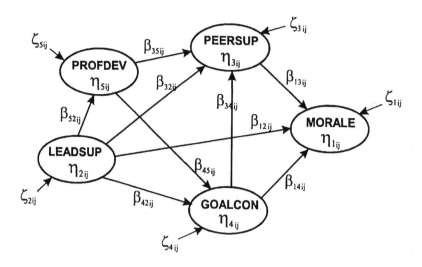

FIG. 12.2. Recursive, two-level SEM for teachers' morale.

levels, and Ψ_{ij} is the variance–covariances among the ζ_{ij}'s. In the typical single-level situation, the measurement model for the observed y_i variables is given by:

$$y_i = \Lambda_{yi}(I - B_i)^{-1}\zeta_i + \varepsilon_i \tag{12.14}$$

where Λ_{yi} is a matrix of factor coefficient loadings of y_i on η_i, and ε_i is a vector of measurement errors for the y_i variables. The covariance matrix of y_i is written as:

$$\Sigma_i = \Lambda_{yi}(I - B_i)^{-1}\Psi_i(I - B_i')^{-1}\Lambda_{yi}' + \Theta_{\varepsilon i} \tag{12.15}$$

where Ψ_i is the variance–covariances among the ζ_i's, and $\Theta_{\varepsilon i}$ is the covariances among the ε_i's.

In the present case, however, the y_i variables are composites such that y_i and η_i are declared to be equivalent (i.e., $y_i \equiv \eta_i$). Moreover, for the two-level case, $y_{ij} \equiv \eta_{ij}$, where the parameters from the measurement model in Equations 12.14 and 12.15, namely Λ_{yi} and $\Theta_{\varepsilon i}$ at the teacher level, were constrained to be invariant at the school level (i.e., $\Lambda_{yj} \equiv \Lambda_{yi}$, and $\Theta_{\varepsilon j} \equiv \Theta_{\varepsilon i}$). Such an approach is important to ensure that the composite constructs have equivalent meanings and interpretations at the two levels.

There are several advantages of using the sub-model described by Equation 12.13. First, it may be preferred because it has only two parameter matrices to be estimated, namely, B_{ij}, and Ψ_{ij}. Of greater importance, there are strong substantive grounds for treating both observed (y) and latent variables (η) as endogenous. Influenced by recent developments in chaos theory, there is a growing body of opinion from within modern systems theory (Simon, 1993) suggesting that all elements within psychosocial systems are endogenous, including a person's gender, socioeconomic status, and so on. In fact, psychosocial theorists and researchers are finding increasing difficulty in justifying the nomination of certain variables as independent or exogenous. Besides, once either an observed (x) or composite variable (ξ_c) has been declared as independent or exogenous, one can neither estimate the regression effects among them, nor estimate path coefficients from endogenous (η) to exogenous (ξ_c) constructs — by definition! For substantive research applications in the use of such models, see Rowe and Rowe (1992a, 1992b, 1999).

Under a maximum likelihood method of estimation, the model shown in Fig. 12.2, and specified by Equation 12.13, was fitted simultaneously to the school- and teacher-level variance–covariance matrices given in Table 12.5. The values of Λ_y and Θ_ε for each composite construct were fixed to be those obtained from fitting the relevant one-factor congeneric measurement models for the total teacher sample, specified by Equations 12.1 and 12.3 and tabulated in the labeled columns of Table 12.1. The relevant LISREL 8.30 input file is available from the author.

To assist interpretation of the obtained parameter estimates from the fitted model, the completely standardized solution is presented in Fig. 12.3, with the school-level estimates in normal type and those for the teacher-level in bold type. For completeness, the standardized direct, indirect, and total effect estimates at the school and teacher levels, are presented in Table 12.6. From Fig. 12.3, all parameter estimates with one exception are statistically significant by univariate two-tailed tests. In the interests of conserving space, further details regarding the solution to this model may be obtained by requesting the relevant LISREL file available from the author.

Comment

The solution to Model 3 (fitted simultaneously at the school and teacher levels) summarized in Fig. 12.3 and Table 12.6 indicate that the obtained goodness-of-fit indices of the model to the data are excellent. A substantive interpretation of the solution suggests at least two notable features of the fitted multilevel, structural equation model. First, the direct effects of goal congruence (goalcon) on morale (morale) and peer support (peersup),

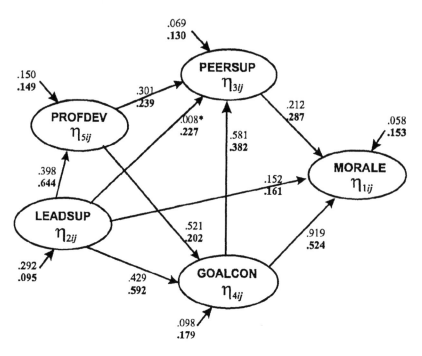

Model Goodness-of-fit Indices: $\chi^2(2) = 3.351, p = 0.187$
RMSEA = 0.020; SRMR = 0.010
GFI = 0.991; PNFI = 0.100
NFI = 1.00; CFI = 1.00
IFI = 1.00; RFI = 0.996

FIG. 12.3. Common metric completely standardized solution to fitted multilevel, structural equation model for morale, showing direct effects and residual variances at the school level (normal type) and teacher level (bold type)

Note: All recorded path coefficients (except those indicated by *) are significant beyond the $p < 0.05$ α level, by univariate two-tailed tests.

TABLE 12.6

Solution to Fitted Multilevel Structural Equation Model Showing Standardized Direct, Indirect, and Total Effects at the School Level (above) and Teacher Level (below)

Leadership Support on	Direct Effect	Indirect Effect	Total Effects
Morale	.125	.569	.694
	.161	.569	.730
Peer support	.009*	.553	.562
	.227	.429	.656
Goal congruence	.457	.221	.678
	.591	.130	.721
Professional development	.486	–	.486
	.643	–	.643

Peer Support on	Direct Effect	Indirect Effect	Total Effects
Morale	.155	–	.155
	.288	–	.288

Goal Congruence on	Direct Effect	Indirect Effect	Total Effects
Morale	.710	.096	8.06
	.526	.110	.636
Peer support	.616	–	.616
	.382	–	.382

Prof'l Development on	Direct Effect	Indirect Effect	Total Effects
Morale	–	.410	.410
	–	.197	.197
Peer support	.279	.280	.559
	.239	.077	.316
Goal congruence	.455	–	.455
	.202	–	.202

* Not statistically significant at the $p < 0.05$ α level by univariate two-tailed test.

and professional development (profdev) on peer support (peersup) and goal congruence (goalcon) are notably stronger at the school level than at the teacher level. In fact, at the school level, the magnitudes of the total effect estimates of professional development (profdev) on morale and goal congruence are more than twice those at the teacher level. Second, the salient feature of the solution is that valuable information at both the teacher and school levels can be obtained by accounting for the hierarchical structure of the data. Furthermore, the evidence from the index of the root mean square error of approximation (RMSEA) suggest a high likelihood in the stability of these interdependent effects in accounting for variation in teachers' morale across additional samples of teachers drawn from similar populations.

CONCLUDING COMMENTS

Using a hierarchically structured data set designed to explain variation in teachers' cognitive/affective constructions of their roles and perceptions of their work environments, the key purpose of this chapter is to draw attention to the importance of accounting for the measurement and distributional characteristics of data typically obtained in psychosocial/ organizational psychological research, and especially their inherent hierarchical structure. Consistent with this aim, the chapter demonstrates the utility of fitting one-factor congeneric measurement models to obtain maximally reliable latent constructs (or composite variables). Following an examination of their distributional properties, the chapter illustrates how such composites may be used in fitting both single-level and multilevel structural equation models to explain variation in response and explanatory variables of interest.

Given the fundamental and enduring contribution of psychosocial inquiry that variations in human cognition, affect, and behavior contribute to and, in turn, are influenced by organizational/contextual factors, it is inappropriate that empirical researchers in the field continue to apply single-level data-analytic methodologies that fail to account for the inherent clustered or hierarchical structure of the data typically obtained. In so doing, substantively important information related to human functioning within social and organizational contexts is not only being denied explication but is being ignored. Moreover, in the absence of the application of analytic methods capable of modeling the structural characteristics of the data, claims to knowledge are at best tentative and at worst spurious.

REFERENCES

Aitkin, M., Anderson, D., & Hinde, J. (1981). Statistical modelling of data on teaching styles. *Journal of the Royal Statistical Society, 144,* (Series A), 419–461.

Alwin, D. F., & Jackson, D. J. (1980). Measurement models for response errors in surveys: Issues and applications. In, D. Schuessler (Ed.), *Sociological methodology.* San Francisco: Jossey-Bass.

Arbuckle, J. L., & Wothke, W. (1999). *AMOS 4.0 users'guide.* Chicago, IL: SPSS, Inc.

Brenner, S. O., Sörbom, D., & Wallius, E. (1985). The stress chain: A longitudinal confirmatory study of teacher stress, coping and social support. *Journal of Occupational Psychology, 58,* 1–13.

Bryk, A. S., & Raudenbush, S. W. (1992). Hierarchical linear models: Applications and data analysis methods. *Advances in quantitative techniques in the social sciences,* (Vol. 1). Beverly Hills: Sage.

Cooper, C. L., & Payne, R. (1992). International perspectives on research into work, well-being, and stress management. In J. C. Quick, L. R. Murphy, & J. J. Hurrell, Jr. (Eds.), *Stress and well-being at work: Assessments and interventions for occupational mental health* (pp. 348–368). Washington, DC: American Psychological Association.

Cronbach, L. J. (1951). Coefficient alpha and the internal structure of tests. *Psychometrika, 16,* 297–334.

Cuttance, P. (1987). Issues and problems in the application of structural equation models. In P. Cuttance & R. Ecob (Eds.), *Structural modeling by example: Applications in educational, sociological and behavioral research* (pp. 241–279). Cambridge: Cambridge University Press.

Cuttance, P., & Ecob, R. (1987). (Eds.). *Structural modeling by example: Applications in educational, sociological and behavioral Research.* Cambridge, Cambridge University Press.

Draper, N. R., & Smith, H. (1981). *Applied regression analysis,* (2nd ed.). New York: Academic Press.

du Toit, M., & du Toit, S. (1999). Multilevel modeling. In K. G. Jöreskog, D. Sörbom, S. du Toit, & M. du Toit, (Eds.), *LISREL 8: New statistical features* (pp. 13–128). Chicago, IL: Scientific Software International Inc.

du Toit, S., du Toit, M., & Cudeck, R. (1999). *Introduction to the analysis of multilevel models with LISREL 8.30.* Chicago, IL: Scientific Software International Inc.

Fleishman, J., & Benson, J. (1987). Using LISREL to evaluate measurement models and scale reliability. *Educational and Psychological Measurement, 47,* 925–939.

Goldstein, H. (1986). Multilevel mixed linear model analysis using iterative generalized least squares. *Biometrika, 73*, 43–56.

Goldstein, H. (1987). *Multilevel models in educational and social research.* London: Griffin.

Goldstein, H. (1995). *Multilevel statistical models.* London: Edward Arnold.

Goldstein, H., & Rasbash, J. (1992). Efficient computational procedures for the estimation of parameters in multilevel models based on iterative generalized least squares. *Computational Statistics and Data Analysis, 13*, 63–71.

Goldstein, H., Rasbash, R., Plewis, I., Draper, D., Browne, W., Yang., M., Woodhouse, G., & Healy, M. (1998). *A user's guide to MlwiN* (Version 1.0). London: Multilevel Models Project, Institute of Education, University of London.

Hart, P. M. (1994). Teacher quality of work life: Integrating work experiences, psychological distress and morale. *Journal of Occupational and Organizational Psychology, 67*, 109–132.

Hart, P. M., Wearing, A. J., Conn, M., Carter, N. L., & Dingle, N. L. (2000). Development of the School Organisational Health Questionnaire: A measure for assessing teacher morale and school organisational climate. *British Journal of Educational Psychology, 70*, 211–228.

Healy, M. J. R., & Goldstein, H. (1976). An approach to the scaling of categorized attributes. *Biometrika, 63*, 219–229.

Hill, P. W., Holmes-Smith, P., & Rowe, K. J. (1993). *School and teacher effectiveness in Victoria: Key findings from Phase 1 of the Victorian Quality Schools Project.* Melbourne: Centre for Applied Educational Research, The University of Melbourne. (ERIC Document Clearing House No. ED 367 067).

Hill, P. W., & Rowe, K. J. (1996). Multilevel modeling in school effectiveness research. *School Effectiveness and School Improvement, 7*, 1–34.

Hill, P. W., & Rowe, K. J. (1998). Modeling student progress in studies of educational effectiveness. *School Effectiveness and School Improvement, 9*, 310–333.

Holmes-Smith, P., & Rowe, K. J. (1994, January). *The development and use of congeneric measurement models in school effectiveness research: Improving the reliability and validity of composite and latent variables for fitting multilevel and structural equation models.* Paper presented at the International Congress for School Effectiveness and Improvement, World Congress Centre, Melbourne.

Hox, J. J. (1994). *Applied multilevel analysis.* Amsterdam: TT-Publikaties.

Jöreskog, K. G. (1971). Statistical analysis of sets of congeneric tests. *Psychometrika, 36,* 109–133.

Jöreskog, K. G. (1990). New developments in LISREL: Analysis of ordinal variables using polychoric correlations and weighted least squares. *Quality and Quantity, 24,* 387–404.

Jöreskog, K. G. (1994). On the estimation of polychoric correlations and their asymptotic covariance matrix. *Psychometrika, 59,* 381–389.

Jöreskog, K. G., & Sörbom, D. (1989). *LISREL 7: A guide to the program and applications.* Chicago: SPSS, Inc.

Jöreskog, K. G., & Sörbom, D. (2000a). *LISREL 8.30: Interactive LISREL for MS Windows 3.1+ Windows '95, Windows '98 and Windows NT.* Chicago, IL: Scientific Software International, Inc.

Jöreskog, K. G., & Sörbom, D. (2000b). *PRELIS 2.30.* In LISREL 8.30: Interactive LISREL for MS Windows 3.1+ Windows '95, Windows '98 and Windows NT. Chicago, IL: Scientific Software International, Inc.

Jöreskog, K. G., Sörbom, D., du Toit, S., & du Toit, M. (1999). *LISREL 8: New statistical features.* Chicago, IL: Scientific Software International Inc.

Kaplan, D., & Elliott, P. R. (1997a). A didactic example of multilevel structural equation modeling. *Structural Equation Modeling, 4,* 1–24.

Kaplan, D., & Elliott, P. R. (1997b). A model-based approach to validating educational indicators using multilevel structural equation modeling. *Journal of Educational and Behavioral Statistics, 22,* 323–347.

Kyriacou, C., & Pratt, J. (1985). Teacher stress and psychoneurotic symptoms. *British Journal of Educational Psychology, 55,* 61–64.

Lawley, D. N., & Maxwell, A. E. (1971). *Factor analysis as a statistical method* (2nd ed.). London: Butterworths.

Marcoulides, G. A., & Schumacker, R. E. (Eds.) (1996). *Advanced structural equation modeling: Issues and techniques.* Mahwah, NJ: Lawrence Erlbaum Associates.

Mardia, K. V., Kent, J. T., & Bibby, J. M. (1979). *Multivariate analysis.* London: Academic Press.

McArdle, J. J., & Hamagami, F. (1996). Multilevel models from a group structural equation perspective. In G. A. Marcoulides & R. E. Schumacker (Eds.), *Advanced structural equation modeling: Issues and techniques* (pp. 89–124). Mahwah, NJ: Lawrence Erlbaum Associates.

McCormick, J., & Solman, R. (1992). Teachers' attributions of responsibility for occupational stress and satisfaction: An organisational perspective. *Educational Studies, 18,* 201–222.

McDonald, R. P. (1985). *Factor analysis and related methods.* Hillsdale, NJ: Lawrence Erlbaum Associates.

McDonald, R. P. (1994). The bilevel reticular action model for path analysis. *Sociological Methods and Research, 22*, 399–413.

McDonald, R. P. (1996). Path analysis with composite variables. *Multivariate Behavioral Research, 31*, 239–270.

McDonald, R. P., & Goldstein, H. (1989). Balanced versus unbalanced designs for linear structural relations in two-level data. *British Journal of Mathematical and Statistical Psychology, 42*, 215–232.

Muthén, B. O. (1994). Multilevel covariance structure analysis. *Sociological Methods and Research, 22*, 376–398.

Rasbash, J., Woodhouse, G., Yang, M., & Goldstein, H. (1996). *MLn: Software for multilevel analysis. Command reference* (Version 1.0a). London: Institute of Education, University of London.

Rasbash, J., Browne, W., Goldstein, H., Yang, M., Plewis, I., Healy, M., Woodhouse, G., Draper, D., Langford, I., & Lewis, T. (2000a). *A user's guide to MLwiN* (Version 2.1). London: Multilevel Models Project, Institute of Education, University of London.

Rasbash, J., Browne, W., Healy, M., Cameron, B., & Charlton, C. (2000b). *MLwiN (Version 1.10.006): Interactive software for multilevel analysis.* London: Multilevel Models Project, Institute of Education, University of London.

Raudenbush, S. W. (1995). Maximum likelihood estimation for unbalanced multilevel covariance structure models via the EM algorithm. *British Journal of Mathematical and Statistical Psychology, 48*, 359–370.

Rowe, K. J. (1989). The commensurability of the general linear model in the context of educational and psychosocial research. *Australian Journal of Education, 33*, 41–52.

Rowe, K. J. (1992, June). *Identifying Type I errors in educational and social research: Comparisons of results from fitting OLS and multilevel regression models to hierarchically structured data.* Paper presented at the Third National Social Research Conference, The University of Western Sydney, Hawkesbury, Australia.

Rowe, K. J. (1995, July). Methodological issues in organizational psychological inquiry: A discussion and worked example. In *Advances in statistical modeling of organizational processes.* Symposium conduted at the Inaugural Industrial and Organizational Psychology Conference, Sydney, Australia.

Rowe, K. J. (1999a). *Multilevel analysis with MLn/MlwiN & LISREL 8.30: An integrated course* (3rd ed.). The 15th ACSPRI Summer Program in Social Research Methods and Research Technology, The Australian National University. Melbourne: Centre for Applied Educational Research, The University of Melbourne.

Rowe, K. J. (1999b). *Advanced structural equation modeling with LIS-*

REL 8.30: A Thematic integrated course (3rd ed.). The 15th ACSPRI Summer Program in Social Research Methods and Research Technology, The Australian National University. Melbourne: Centre for Applied Educational Research, The University of Melbourne.

Rowe, K. J. (2001). *Multilevel structural equation modeling with MLn/MLwiN & LISREL8.30: An integrated course* (5 ed., ISBN 0 86431 357 9). The 17th ACSPRI Summer Program in Social Research Methods and Research Technology, The Australian National University. Camberwell, Vic: The Australian Council for Educational Research.

Rowe, K. J., & Hill, P. W. (1998). Modeling educational effectiveness in classrooms: The use of multilevel structural equations to model students' progress. *Educational Research and Evaluation, 4* (4), 307–347.

Rowe, K. J., Hill, P. W., & Holmes-Smith, P. (1995). Methodological issues in educational performance and school effectiveness research: A discussion with worked examples (leading article). *Australian Journal of Education, 39,* 217–248.

Rowe, K. J., & Rowe, K. S. (1992a). The relationship between inattentiveness in the classroom and reading achievement (Part A): Methodological issues. *Journal of the American Academy of Child and Adolescent Psychiatry, 31* (2), 349–356.

Rowe, K. J., & Rowe, K. S. (1992b). The relationship between inattentiveness in the classroom and reading achievement (Part B): An explanatory study. *Journal of the American Academy of Child and Adolescent Psychiatry, 31* (2), 357-368.

Rowe, K. J., & Rowe, K. S. (1999). Investigating the relationship between students' attentive–inattentive behaviors in the classroom and their literacy progress. In *International Journal of Educational Research, 31* (1-2), 1–137.

Schumacker, R. E., & Lomax, R. G. (1996). *A beginner's guide to structural equation modeling.* Mahwah, NJ: Lawrence Erlbaum Associates.

Simon, H. (Ed.) (1993). *Developments in systems theory: Present dilemmas and future prospects.* New York: McGraw-Hill.

Steiger, J. H. (1999). SEPATH Module. In *STATISTICA for Windows, computer software manual.* Tulsa, OK: StatSoft, Inc.

Werts, C. E., Rock, D. R., Linn, R. L., & Jöreskog, K. G. (1978). A general method for estimating the reliability of a composite. *Educational and Psychological Measurement, 38,* 933–938.

Williams, L. J. (1995). Covariance structure modeling in organizational research: Problems with the method versus applications of the method. *Journal of Organizational Behavior, 16,* 225–233.

Woodhouse, G. (Ed.). (1996). *Multilevel modelling applications: A guide for users of MLn.* London: Multilevel Models Project, Institute of Education, University of London.

13

Design Issues in Multilevel Studies

Steven P. Reise
Naihua Duan
University of California, Los Angeles

Statistical techniques for the analysis of nested data structures are increasing in popularity (Bryk & Raudenbush, 1992; Goldstein, 1995; Kreft & de Leeuw, 1998; Little, Schnabel, & Baumert, 2000). These techniques, which we refer to as multilevel modeling (MLM), have been proposed as a general data-analytic strategy for the exploration of longitudinal data, the effects of community-based interventions, the outcomes of clinical trials, effect sizes in meta-analysis, school effectiveness research, and family process or couples research.

The current popularity of MLM strategies derives in part from their ability to overcome the limitations of traditional approaches (e.g., mixed-effects ANOVA) used in the analysis of nested data structures (Searle, 1987; Zucker, 1990). Although statistical and computational advances in MLM are occurring rapidly (see, for example, the chapters in the present volume), limited attention has been paid to the many complex design issues that are inherent in multilevel studies. The goal of this chapter is to raise and discuss several of these design issues.

The following discussion is divided into three sections. In the first, we examine sample design issues in multilevel studies. Here we emphasize the importance of representative sampling and how the choice of sampling plan depends on the study purpose. We also discuss optimal design, the choice of estimand (house vs. senate model), and cohort versus repeated cross-sectional sampling in longitudinal studies. In the second section, we examine experimental design issues. These topics include the level of randomization, split plot designs, and factorial designs. Finally, in the third section, we discuss the estimation of power.

SAMPLE DESIGN FOR MULTILEVEL STUDIES

Multilevel studies impose important challenges in the sample design. Most multilevel studies are multi-stage in nature: We sample school districts, then sample schools within the sampled school districts, then sample classrooms within the sampled schools, then sample students within the sampled classrooms. In order to make valid inference about the target populations of interest — all school districts in the United States, all public schools, all classrooms, all students, the sample needs to be designed and recruited meticulously to guarantee appropriate representation at all levels.

We strongly urge investigators conducting multilevel studies to use the best efforts possible to obtain representative samples at all levels in order to maximize the validity of the findings. For an example of such a study, we refer readers to Frankel et al. (1999) for a description of the HIV cost and services utilization study, a three-level study with HIV patients in care nested in providers, and providers nested in geographical locales, to provide a nationally representative sample of those patients.

It is understandable that rigorous probability sampling might be impractical or prohibitively costly for many multilevel studies, therefore the investigator might need to resort to a less representative sampling method such as quota sampling or convenience sampling. We urge the investigators for those studies to clarify the sampling method used, and to discuss the implications on the external validity of the study.

The appropriate sample design depends on the purpose of the study. Although we made a general recommendation that investigators should make an effort to obtain a representative sample at all levels, there are exceptions to this recommendation for multilevel studies focused on the group level. Consider, for example, a two-level study of patients nested in providers. If the focus of the study is on patient outcomes, we should attempt to obtain a representative sample on both the provider level and the patient level. If the focus of the study is on providers' skill, the role of the patients might be analogous to test items in a skill test. More specifically, the patients' outcomes are used as gauges for their providers' skills. For this type of study, it would still be important to obtain a representative sample of providers, but it might be more appropriate to take a "standardized sample" of patients across providers with similar compositions in terms of patient characteristics, analogous to the use of standardized tests. For example, if one provider usually sees patients with less severe conditions, the average outcomes for his patients might overstate his skill. Instead, we should oversample his patients with more severe conditions, so that he is compared to other providers on more comparable "test items".[1]

SAMPLE ALLOCATION ACROSS LEVELS

An important decision in designing multilevel studies is how to allocate the sample across levels: Should we take more schools and fewer students in each sampled school, or vice versa? We need to balance between two counterbalancing factors. On the one hand, the possible presence of intracluster correlation usually indicates that for the same total sample size of individuals, it is better to spread them across as many groups as possible to reduce the design effect due to clustering. In other words, we should use small cluster sizes. On the other hand, for the same total sample size of individuals, the cost of the study usually increases with the number of groups: There is usually some cost to recruit each school and obtain school-level measures, in addition to the cost to recruit and measure the students. Therefore, it is better to use large cluster sizes, to take as many students as possible in the same school to spread out the school level costs. The challenge to the multilevel study is therefore how to balance between the two opposing forces.

Snijders and Bosker (1993) provided equations for computing the optimal number of groups and subjects within groups to achieve desired standard errors for the subject- and group-level coefficients in MLM. For a two-level study, let us assume that the study cost is linear in both the number of groups, k, and the number individuals per group, n,

$$TotalCost = k * (c_g + c_s * n), \tag{13.1}$$

where c_g denotes the cost for recruiting each group, and c_s denotes the cost for recruiting each individual subject after having recruited the group. We need to determine the optimal cluster size, n, and the optimal number of groups, k.

The optimal sample allocation for estimating the population mean is achieved by setting the cluster size as

---

[1]The heterogeneity in patient composition can be handled through case mix adjustment, such as the use of standardized mortality rate that adjusts the mortality rate by patient age, gender, and maybe other demographic characteristics. This is necessary when the sample of patients available from each provider is not standardized. The use of a standardized sample of patients can reduce or eliminate the need to adjust for case mix in the analysis stage, and can also improve the precision and power for the analysis.

$$n(optimal) = \sqrt{\frac{((1-\rho)/\rho)}{(c_g/c_s)}}, \qquad (13.2)$$

where ρ denotes the intracluster correlation coefficient (Cochran, 1977; Snijders & Bosker, 1993). The optimal cluster size given in Equation 13.2 is optimal in the sense of minimizing the standard error for the estimated population mean under a fixed total cost; it is also optimal in the sense of minimizing the total cost under the constraint of a fixed, prespecified standard error.

The optimal group size in Equation 13.2 does not depend on the number of groups, k, thus the two components of the design (n and k) can be determined separately. This is a convenient feature for planning a study. The investigator can first assess the optimal group size (assuming that he has data or prior knowledge to assess the relative cost ratio c_g/c_s, and the intraclass correlation ρ), then conduct power calculations to determine the number of groups.

It should be noted, though, that Equation 13.2 depends on the assumption in Equation 13.1 that the cost per group and the cost per subject are both constant. If those assumptions are not met, the optimal group size might vary with the number of groups. This makes it more difficult to design the study, such as to conduct the power calculations if both the number of groups and the group size need to be considered simultaneously. Furthermore, Equation 13.2 also assumes that the intracluster correlation coefficient stays constant across groups. For intervention studies, one of the by-products of the intervention might be to increase the similarity of subjects within groups in the treatment condition relative to controls, thus the intracluster correlation coefficient might differ across groups, further complicating the design of the study.

Raudenbush (1997) considered the effects of subject-level covariates in the context of optimal design in MLM contexts. As he pointed out, optimal designs for clustered sampling depend on the intraclass correlation, and the cost of sampling at the various levels (see Equation 13.1). He also pointed out that in many studies, there are covariates available that can be included in the model to decrease between-group variation. The optimal cluster size n is larger and the corresponding k is smaller when covariates are used than when covariates are not used. Further extensions of those results to three-level models and logistic regression are given in Moerbeek (2000).

Extending this line of research, Raudenbush and Liu (2000) developed effect size indices for treatment effects and treatment-by-site variance terms.

In turn, they described how optimal allocation of resources between persons and sites depends on whether the researcher is interested in estimating main effects or moderator effects.

CHOICE OF ESTIMAND: SENATE MODEL VS. HOUSE MODEL

An important question that deserves careful consideration in multilevel studies is what estimand we should attempt to estimate. Let us consider one of the most prominent multilevel designs that was established more than 200 years ago, the design of the U.S. Congress. We have individuals nested in states in this multilevel design problem. It is intriguing that our founding fathers adopted two distinct designs, the Senate and the House of Representatives. The Senate provides an unweighted average across the states. The House provides a weighted average across the states, weighted by each state's population.   The two designs each address a different question: The Senate is meant to represent the states' interests, the House is meant to represent the interests of the general population. As history has shown, the two interests very often converge, but they do diverge on many important occasions.

The same distinction between the Senate model and the House model deserves to be considered in many multilevel studies.   The usual specification of multilevel models implicitly assumes the Senate model: The group level model usually models the groups as exchangeable entities (after controlling for group level fixed effects). For example, in a study of students nested in states, we might examine the impact of school funding mechanism on student outcomes, such as whether states with a centralized funding mechanism have better student outcomes. However, states vary in size. If we average state-level outcomes across all states, the influence of each state would be the same irrespective of the number of students in each state. Whereas this Senate model characterization of states might be appropriate for certain research questions focused on states, it is inappropriate for many research questions that are focused on students. Therefore, it is important to entertain the House model as an alternative to the Senate model in many multilevel studies, and weight the states by the enrollment in the group-level analyses.

More specifically, the question is what estimand should we attempt to estimate?   Consider the scenario that we have a complete census of all states with centralized funding, all students in those states, and all of their outcomes. Let Y_{ij} denote the outcome for the j-th student in the i-th state, N_i denote the total number of students in the i-th state, and μ_i denote the expected outcome for the i-th state,

$$\mu_i = E(Y_{ij}|i) = \Sigma_j Y_{ij}/N_i. \tag{13.3}$$

The usual specification for the group level model,

$$\mu = E(\mu_i) = \Sigma_i \mu_i / K, \tag{13.4}$$

where K denotes the number of states with centralized funding, μ_g specifies the unweighted mean across states as the estimand, implicitly following the Senate model. However, if the objective of the study is to improve student outcomes, it would be more appropriate to specify the weighted mean across states as the estimand, following the House model:

$$\mu^* = E^*(\mu_i) = \Sigma_i(N_i\mu_i)/\Sigma_i N_i \tag{13.5}$$

where E^* denotes taking the expectation across the states weighted by the total number of students in each state.

COHORT DESIGNS VERSUS REPEATED CROSS-SECTIONAL DESIGNS

In any relatively long-term study, a researcher has the option of selecting a cohort design, where individuals are followed across the study period, or a repeated cross-sectional design, where new individuals are sampled at each study measurement point. For simplicity of discussion, we are ignoring the possibility of mixtures of these two extremes.

For the same number of measurements,[2] cohort studies are usually more precise than repeated cross-sectional studies for estimating changes over time because they control for individual-level effects through comparisons of repeated observations on the same individuals. On the other hand, repeated cross-sectional studies are usually more precise for estimating mean levels (e.g., the prevalence of a rare disease) because they include more individuals, thus reducing the sampling error on the individual level.

The cost for the two types of designs usually differ. With the cohort design, the initial recruitment cost incurs only once for each individual. With the repeated cross-sectional design, the initial recruitment cost incurs in every wave. Cohort designs, on the other hand, require costs of tracking individuals and are subject to attrition and bias (see Little, Lindenberger, & Maier, 2000). As time goes by, the cohort sample may become less representative of the target population. Compton, California, was once predominantly African American but now has a large Latino population. It provides an excellent example. For this reason, some cohort studies replenish the sample to include new members in the target population.

---

[2]Counting individuals in cohort studies multiple times by the number of repeated measurements obtained.

On the other hand, the cross-sectional design may have exactly the opposite problem, namely, successive samples may differ significantly on important covariates. The choice of longitudinal sampling design is a complex choice. Feldman and McKinlay (1994) present a statistical framework for balancing the issues of sample size, precision, and bias in deciding between these competing designs. As pointed out in Murray et al. (1994, p. 496), to determine the relative efficiency of a cohort versus a cross-sectional design, researchers need estimates of relevant parameters such as variance components and autocorrelations.

EXPERIMENTAL DESIGN FOR MULTILEVEL STUDIES

A major application of multilevel modeling is in the evaluation of social, psychological, or educational interventions, such as multi-site randomized trials with patients nested in sites. The sample design issues discussed in the previous section also apply to experimental studies: In order for the experimental studies to be generalizable to the target population of interest, it is essential that the sample on which the experimental study is conducted be representative of the target population. In addition, experimental studies impose additional design issues.

Level of Randomization

Unlike single-level studies, multilevel studies often offer a choice of which level should be used to randomize the experimental treatments. Should we randomize at the school level, assigning all classrooms and students in the same school to the same treatment? Should we randomize at the classroom level, assigning some classrooms in each school to each treatment condition, but holding all students in the same classroom to the same treatment condition? Or should we randomize at the student level, assigning some students in each classroom to each treatment condition?

It should be noted that some treatment conditions are delivered at a group level, which prohibits the experimental assignment from being randomized at a lower level. For example, if we study the effect of a training program for teachers, it is usually anticipated that the impact of the treatment will influence all students in the classrooms of the teachers who received the training. Therefore, we cannot randomize the treatment at the student level — it is practically impossible, and rather unethical, to ask the teacher to apply the skills learned from the training program only on some students in her classroom, and withhold the skills from the other students.

In terms of statistical precision, it is usually advantageous to randomize at the lowest level possible, to reduce the design effect due to clustering. For the training program discussed earlier, this would imply randomizing at the classroom level. If the study is conducted in multiple schools, this design will allow us to minimize the design effect due to clustering at the school level. The main effects of schools is cancelled out when we compare classrooms assigned to the training program versus classrooms assigned not to receive the training program.

However, contamination might be a concern under this design. Although the teachers are randomized to the training programs, it is conceivable that they might share the skills they learned from the program, thus exposing the teachers assigned not to receive the training program (and their students) to the training program indirectly. If such contamination occurs, the intervention effect would be underestimated.

If contamination is a major concern, an alternative design is to randomize at a higher level, say, at the school level, to reduce the likelihood or the level of contamination. However, the investigator needs to balance the trade-off between contamination and the clustering effect: Randomizing at the school level can reduce the power and precision of the study, for the same sample size at each level, due to the clustering effect at the school level.

The same sample allocation issues discussed earlier also apply to experimental studies. We refer interested readers to Raudenbush (1997), Raudenbush and Liu (2000), and Moerbeek (2000) for detailed discussions.

Split Plot Design and Factorial Design

We assumed implicitly in the previous section that the experimental study has to randomize at a single level. For many multilevel experimental studies, there is a potential to use the split plot design and randomize various components of the intervention at different levels. For example, consider a hypothetical intervention that includes both a training program for teachers and an individual tutoring program for students to be delivered by "big brothers" and "big sisters" from upper grade classes. Because the training program is delivered at the classroom level, it appears that we have to randomize the intervention at the classroom level (or a higher level, if contamination is a concern). However, an alternative that might provide more useful information is to use a split plot design and randomize the training program at the classroom level, and randomize the tutoring program at the student level. (We assume here that the tutoring program and the teacher training program are not directly related; in particular, a student can participate in the tutoring program whether or not her teacher participated in the training program.) This design allows us to estimate the effects separately for the two program components, and the interaction between them. Furthermore, randomizing the tutoring program at the individual level provides better precision for estimating the main effect of the tutoring program.

The split plot design is an example of a factorial experiment that tests multiple intervention components within the same study. Those experimental designs are used less often in behavioral intervention studies (whether single level or multilevel) than in agricultural and engineering studies. We believe there is a promising potential and that factorial experimental designs should be applied more broadly in behavioral intervention studies, especially for multilevel studies, many of which involve multifaceted interventions at multiple levels.

The experimental design commonly used in behavioral intervention studies can be characterized as a "kitchen sink design." That is, we usually include all promising intervention components into the intervention, and compare with a control condition that excludes all intervention components. Such a kitchen sink design might be optimally powered[3] for a first-generation study to prove the concept for the study. However, the study finding is usually limited to whether the intervention (as a bundle) is effective; beyond this basic question, we know very little about which intervention components are essential, which can be eliminated, and how the components work together. Therefore, we believe it is important for investigators to move beyond the first-generation kitchen sink design and use factorial experimental designs to "tease apart" the effects of various intervention components.

Power Analysis for Multilevel Studies

In most multisite trials, groups (e.g., classrooms, clinics), not individuals, are randomly assigned to treatments. This design usually leads to a clustering effect — a positive intraclass coefficient that typically lowers power. Numerous researchers have shown that ignoring clustering can lead to serious errors in interpreting the results of statistical significance tests. Specifically, when there is non-independence among observations that is

---

[3]As noted earlier, the power and precision might be reduced due to clustering if the bundled intervention has to be delivered at a higher level. It is possible that the power and precision might be higher for a stripped down intervention that can be delivered at a lower level. For example, for the same sample size at all levels, the power and precision might be higher for testing the tutoring program alone at the student level than testing the tutoring program and the teacher training program jointly at the classroom level. In other words, the kitchen sink design might not be optimally powered even for a first-generation study.

not accounted for in the model, then standard errors can be seriously underestimated and the Type I error rate is thus inflated (Aitkin, Anderson, & Hinde, 1981; Barcikowski, 1981; Blair, Higgins, Topping, & Mortimer, 1983).

Consider the linear random effects model shown in Equation 13.6. In this model, Y_{ij} is the dependent variable score of subject j in group i, μ is the grand mean, α_i is the group effect, and ε_{ij} is the subject error.

$$Y_{ij} = \mu + \alpha_j + \varepsilon_{ij}. \qquad (13.6)$$

The variances of the group and individual effects can be symbolized by σ_α^2 and σ_ε^2, respectively. Given this formulation, then the intraclass correlation is $\rho = \sigma_\alpha^2 \,/\, (\sigma_\alpha^2 + \sigma_\varepsilon^2)$.

To explain the effects of clustering more fully, we need to introduce the concept of the design effect (DE) or the variance inflation factor (Kish, 1965). The design effect is shown in Equation 13.7. The term in the numerator indicates the real error variance and the term in the denominator, σ^2/n,

$$DE = Var(\hat{\mu})/(\sigma^2/n) \qquad (13.7)$$

represents the least squares estimate of the error variance. This ratio can be approximated by Equation 8 where m* is the "typical" cluster (group) size.

$$1 + (m^* - 1)\rho. \qquad (13.8)$$

Clearly, if ρ is zero (no clustering effect), then there is no design effect and all estimates of error variance are appropriate. However, as ρ goes toward 1.0, then DE becomes large and the degree of overestimated precision becomes substantial. It is also clear from Equation 13.8 that as the group sizes increase, the effects of clustering also increase. Finally, the design effect will be increased to the degree to which there is not an equal number of subjects per group.

To illustrate the effects of clustering, consider a simple experiment where there are 10 groups, 5 of which have been assigned to a treatment condition and 5 to a control condition. Assume also that there are 20 subjects per group, so the total sample size is 200 subjects (100 treatment and 100 control). Because of clustering effect, we can no longer use the nominal sample size in our power computation for testing a treatment versus control difference, rather we need to consider the "effective" sample size given the design effect defined as the nominal sample size divided by the design effect.

For example, if $\rho = 0.05$, then the design effect is 1.95, and the effective sample size is $100/1.95 = 51.28$ subjects per treatment conditions; if ρ

= 0.10, then the design effect is 2.90, and the effective sample size is 100/2.90 = 34.48 subjects per treatment conditions. Thus, in the presence of clustering effect, subjects do not provide unique information. In turn, the effective sample size can be significantly lower than the nominal sample size. This negatively impacts our power to detect the treatment versus control group difference.

Power issues in MLM are complex due to the intraclass correlation. If the intraclass correlation is zero, then we could use the standard power analysis methods based on single-level models. However, with nested designs, clustering is the rule, not the exception. Barcikowski (1981) provided tables that detail how the power of testing group-level mean differences is affected by clustering and the combination of the number of groups and the number of persons per group. Brown and Liao (1999) provided a reference to online power calculators for three-level models in school-based research, and for latent growth modeling. Muthén and Curran (1997) and Hedeker, Gibbons, and Waternaux (1999, p.   90) also provided resources and programs for computing power in longitudinal designs.   Finally, Raudenbush and Liu (2000) cited an optimal design (see following discussion) power calculation software package available from Raudenbush.

The new power calculators are a great addition to the applied researcher's repertoire. Nevertheless, power questions in multilevel designs remain unwieldy and underresearched, especially under conditions of a small number of persons within groups and under different estimation methods (e.g., empirical Bayes). The chief problem, however, is the inherent complexity of multilevel studies.   In any reasonably complex model, a researcher may be interested in the power to detect moderator or mediator effects (Krull & MacKinnon, 2001), the size of variance components (Raudenbush & Liu, 2000), or the impact of individual-level (time-varying) or group-level (time-invariant) covariates.   Kreft and de Leeuw (1998, pp. 119–126) reviewed some of these complexities and observed that few empirical investigations of these issues are published studies.

A further complexity with any power or optimal design analyses is the effects of covariates.   As is well known in ordinary regression, the power to detect an effect (e.g., a partial r) can be raised considerably by including the proper covariates into the model. The situation is even more challenging in multilevel level modeling contexts because there are typically potential covariates at more than one level (e.g., subject and group level, or time-varying and time-invariant covariates in longitudinal studies). Determining the precise impact of a covariate on power a priori is complicated in multilevel studies because it depends on a number of factors, including how much within and between variance a covariate accounts for

(Raudenbush & Liu, 2000).

SUMMARY

It can be said that an analysis is only as good as the data collected. Garbage in, garbage out; the most elegant analysis cannot make up for the flaws in the study design. This rubric is especially true in multilevel studies where researchers need to consider sampling, power, and external validity issues at multiple levels of inference, and must confront the problem of clustering. Compared to single-level designs, MLM calls for more decisions and more care in the design. These decisions need to be informed.

Perhaps the prevailing theme of this chapter is that new tools are being developed that allow researchers to make more informed decisions. Given the often tremendous costs of multilevel studies, these new tools are most welcome. For example, in the context of optimal design and power calculation, we cited recently developed computer programs. There is no longer any excuse for the usual practice of treating a multilevel study as if it were a single-level study for purposes of estimating power. For another example, in considering the choice between a cross-sectional or cohort design, we cited new statistical developments that allow researchers to more precisely define the relative advantages of each.

REFERENCES

Aitkin, M. (1999). A general maximum likelihood analysis of variance components in generalised linear models. *Biometrics, 55,* 117–128.

Aitkin, M., Anderson, D., & Hinde, J. (1981). Statistical modeling of data on teaching styles. *Journal of the Royal Statistics Society, (144, Series A),* 419–461.

Barcikowski, R. S. (1981). Statistical power with group mean as the unit of analysis. *Journal of Educational Statistics, 6,* 267–285.

Blair, R. C., Higgins, J. J., Topping, M. E., & Mortimer, A. L. (1983). An investigation of the robustness of the *t* test to unit of analysis violations. *Educational and Psychological Measurement, 43,* 69–80.

Brown, C. H., & Liao, J. (1999). Principles for designing randomized preventive trials in mental health: An emerging developmental epidemiology paradigm. *American Journal of Community Psychology, 27,* 673–709.

Bryk, A. S., & Raudenbush, S. W. (1992). *Hierarchical linear models.* Newbury Park, CA: Sage.

Cochran, W. G. (1977). *Sampling techniques* (3rd ed.). New York: Wiley.

Feldman, H. A., & McKinlay, S. M. (1994). Cohort versus cross-sectional design in large field trials: Precision, sample size, and a unifying model. *Statistics in Medicine, 13*, 61–78.

Frankel, M. R., Shapiro, M. F., Duan, N., Morton, S. C., Berry, S. H., Brown, J. A., Burnam, M. A., Cohn, S. E., Goldman, D. P., McCaffrey, D. F., Smith, S. M., St Clair, P. A., Tebow, J. F., & Bozzette, S. A. (1999). National probability samples in studies of low-prevalence diseases. Part II: Designing and implementing the HIV cost and services utilization study sample. *Health Services Research, 34* (5), 969–992.

Goldstein, H. (1995). *Multilevel statistical models* (2nd ed.). New York: Halsted.

Hedeker, D., Gibbons, R. D., & Waternaux, C. (1999). Sample size estimation for longitudinal designs with attrition: Comparing time-related contrasts between two groups. *Journal of Educational and Behavioral Statistics, 24*, 70–93.

Kish, L. (1965). *Survey sampling.* New York: Wiley.

Kreft, I., & de Leeuw, J. (1998). *Introducing multilevel modeling.* London: Sage.

Krull, J. L., & MacKinnon, D. P. (2001). Multilevel modeling of individual and group level mediated effects. *Multivariate Behavioral Research, 36*, 249–278.

Little, T. D., Lindenberger, U., & Maier, H. (2000). Selectivity and generalizability in longitudinal research: On the effects of continuers and dropouts. In, T. D. Little, K. U. Schnable, & J. Baumert (Eds.), *Modeling longitudinal and multilevel data: Practical issues, applied approaches, and specific examples* (pp. 287–300). Mahwah, NJ: Lawrence Erlbaum Associates.

Little, T. D., Schnabel, K. U., & Baumert, J. (2000). *Modeling longitudinal and multilevel data: Practical issues, applied approaches, and specific examples.* Mahwah, NJ: Lawrence Erlbaum Associates.

Moerbeek, M. (2000). *Design and analysis of multilevel intervention studies.* Unpublished doctoral dissertation, Department of Methodology and Statistics, Maastricht University, Maastricht, The Netherlands.

Murray, D. M., McKinlay, S. M., Martin, D., Donner, A. P., Dwyer, J. H., Raudenbush, S. W., & Graubard, B. I. (1994). Design and analysis issues in community trials. *Evaluation Review, 18*, 493–514.

Muthén, B. O., & Curran, P. J. (1997). General longitudinal modeling of individual differences in experimental designs: A latent variable framework for analysis and power estimation. *Psychological Methods, 2*, 371–402.

Raudenbush, S. W. (1997). Statistical analysis and optimal design for cluster randomized trials. *Psychological Methods, 2,* 173–185.

Raudenbush, S. W., & Liu, X. (2000). Statistical power and optimal design for multisite randomized trials. *Psychological Methods, 5,* 199–213.

Searle, S. R. (1987). *Linear models for unbalanced data.* New York: Wiley.

Snijders, T. A. B., & Bosker, R. J. (1993). Standard errors and sample sizes for two-level research. *Journal of Educational Statistics, 18,* 237–259.

Zucker, D. M. (1990). An analysis of variance pitfall: The fixed effects analysis in a nested design. *Educational and Psychological Measurement, 50,* 731–738.

Subject Index

Author Index

Author Information

Bachmann, Nicloe. Dept. for Public Health and Social Welfare, Rathausgasse 1, CH-3011. Bern, Switzerland.

Baumler, Elizabeth Razniak. University of Texas–Houston Health Science Center, 3419 East Cedar Hollow, Pearland, TX 77584.

Bentler, Peter M. University of California, Los Angeles, Departments of Psychology and Statistics, Box 951563, Los Angeles, CA 90095-1563.

Cudeck, Robert. Professor, University of Minnesota, Psychology Department, 75 East River Rd., Minneapolis, MN 55455. Webpage: www.psych.umn.edu/psyfac/core/Psychometrics/cudeck.htm.

de Leeuw, Edith D. Survey methodologist/survey statistician, MethodikA, Plantage Doklaan 40 NL-1018 CN Amsterdam. Webpage: http://www.xs4all.nl/ edithL.

Duan, Naihua. Professor in Residence, Center for Community Health, UCLA Neuropsychiatric Institute, UCLA Wilshire Center, UCLA Campus mail code 705146, 10920 Wilshire Blvd. Suite 350, Los Angeles, CA 90024-6521.

du Toit, Stephen H. C. Manager of Technical Operations, Scientific Software International, Inc., 7383 N. Lincoln Avenue, Suite 100, Lincolnwood, IL 60712-1704.

Ecob, Russel. Director of Ecob Consulting. Ecob Consulting, 36 Prospecthill Road, Glasgow G42 9LE, Scotland, England. Webpage: http:www.ecob-consulting.com.

Fielding, Antony. Senior Lecturer in Social Statistics, University of Birmingham, Department of Economics, Birmingham B15 2TT, England. Webpage: http://www.bham.ac.uk/economics/staff/tony.htm.

Hox, Joop J. Professor, Utrecht University , Department of Methodology & Statistics, P.O. Box 80140, NL-3508 TC Utrecht, the Netherlands. Webpage: http://www.fss.uu.nl/ms/jh.

Hutchison, Dougal. Cheif Statistician, National Foundation for Educational Research, The Mere, Upton Park, Slough SL1 2DQ, England. Webpage: http://www.nfer.ac.uk.

Jo, Booil. Postdoctoral Research Associate, University of California, Los Angeles, Graduate School of Education & Information Studies, Social Research Methodology Division, Los Angeles, CA 90095-1521. Email: booil@ucla.edu.

Muthén, Bengt. Professor. University of California, Los Angeles, Graduate School of Education & Information Studies, Moore Hall, Box 951521, Los Angeles, CA 90095-1521.

Reise, Steven P. Professor. University of California, Los Angeles, Department of Psychology, Los Angeles, CA 90095-1563.

Rowe, Ken.   Principal Research Fellow, Australian Council for Educational Research, 19 Prospect Hill Road, Camberwell, VIC 3124, Australia. Webpage: http://www.acer.edu.au.

Seltzer, Mike.   Associate Professor, University of California, Los Angeles, Graduate School of Education & Information Studies, Moore Hall, Box 951521, Los Angeles, CA 90095-1521. Webpage: http://www.gseis.ucla.edu/faculty/pages/seltzer.html.